Information Security and Cryptography

Information Security – protecting information in potentially hostile environments – is a crucial factor in the growth of information-based processes in industry, business, and administration. Cryptography is a key technology for achieving information security in communications, computer systems, electronic commerce, and in the emerging information society.

Springer's Information Security & Cryptography (IS&C) book series covers all relevant topics, ranging from theory to advanced applications. The intended audience includes students, researchers and practitioners.

More information about this series at http://www.springer.com/series/4752

Yan Lin

Novel Techniques in Recovering, Embedding, and Enforcing Policies for Control-Flow Integrity

Yan Lin
School of Computing and Information
Systems
Singapore Management University
Singapore, Singapore

ISSN 1619-7100   ISSN 2197-845X   (electronic)
Information Security and Cryptography
ISBN 978-3-030-73140-3   ISBN 978-3-030-73141-0   (eBook)
https://doi.org/10.1007/978-3-030-73141-0

This Springer imprint is published by the registered company Springer Nature Switzerland AG
The registered company address is: Gewerbestrasse 11, 6330 Cham, Switzerland

Preface

This book is an introduction on how to make Control-Flow Integrity (CFI) fine-grained, practical, and efficient. CFI is an attractive security property with which most injected and code-reuse attacks can be defeated. Most CFI approaches use a coarse-grained policy, rely on memory page protection mechanism, and result in a large runtime overhead. This book is for those who want to explore more in the world of cybersecurity and want to have a better understanding of CFI.

An important aim of the book is to make readers have a basic concept on what CFI is, why we need it, and the issues for current CFI enforcements. The book is written and presented in a way that is directly accessible to all kinds of readers, no matter you are experts or freshmen in cybersecurity.

The book provides an overview of what CFI is and the issues of current enforcements. The readers can have a detailed understanding on why CFI is important in cybersecurity compared to other security enforcements. Each chapter discusses one possible issue of current CFI implementation and presents the corresponding solution. The reader can get more details about how CFI works from them and broaden their mind to find new solutions to mitigate these limitations.

Singapore
Yan Lin

Contents

List of Figures

List of Tables

Chapter 1
Introduction

In this chapter, we first introduce concepts and implementations of Control-Flow Integrity [1], which is a fundamental approach to mitigating control-flow hijacking attacks, and then present practical issues of previous CFI systems and summarize how we address those problems.

1.1 Overview of Control-Flow Integrity

Application is often written in memory-unsafe languages; this makes it prone to memory errors that are the primary attack vector to subvert systems. Many protection mechanisms including DEP (Data Execution Prevention [2]), ASLR (Address Space Layout Randomization [3]), and GS/SSP (Stack Smashing Protector [4]) have gained wide adoption, and they are making it more difficult for attackers to exploit vulnerabilities. These mechanisms can mitigate various standard attacks, but these reactive defenses can often be bypassed by advanced exploitation techniques [5, 6].

Natural protection against control-flow hijacking attacks is to enforce CFI (Control-Flow Integrity [1]). The goal of CFI is to restrict the set of possible control-flow transfers to those that are strictly required for correct program execution. This prevents control-flow hijacking attacks such as Return-Oriented Programming (ROP) [7–9] from working because they would cause the program to execute control-flow transfers, which are illegal under CFI. Figure 1.1 shows a high-level representation of CFI: first, a Control-Flow Graph (CFG), which approximates the set of legitimate control-flow transfers is construct prior to program execution [10–13]. Next, a CFI check is inserted for indirect branches (e.g., indirect calls, indirect jumps, and returns). These checks ensure that all executed branches correspond to edges in the CFG at runtime. For instance, the valid targets of node 3 can only be either 5 or 6. If the adversary aims to redirect execution to node 4, CFI will immediately terminate the program execution.

© The Author(s), under exclusive license to Springer Nature Switzerland AG 2021
Y. Lin, *Novel Techniques in Recovering, Embedding, and Enforcing Policies for Control-Flow Integrity*, Information Security and Cryptography,
https://doi.org/10.1007/978-3-030-73141-0_1

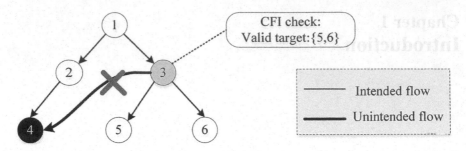

Fig. 1.1 Example of control-flow integrity

Despite CFI's efficacy, it has not seen wide adoption. We believe that not well supporting some critical features contributes to CFI's poor deployment.

First of all, having an accurate CFI policy (CFG) is known to be hard as it is generally difficult to identify the target locations for all control-flow transfers. Most binary-level CFI techniques [1, 10, 11] have to conservatively consider all functions as potential targets of an indirect caller, resulting in loosened CFI policies which make them vulnerable to various attacks [14–16]. Most fine-grained approaches require the availability of the source code [12, 13, 17], so that they can construct a more precise CFG by making use of type information available in the source code. This kind of information is not available in binary since compilers do not preserve much language-level information in the process of compilation. Two fine-grained approaches TypeArmor [18] and τCFI [19] are proposed to recover a fine-grained CFG by matching the function signature (the number of arguments and argument width) at indirect caller and callee sites at the binary level. However, they rely on strictly following the calling convention used by compilers, and various compiler optimizations may violate the calling conventions. For example, modern compilers typically do not (re)set the argument registers explicitly at the caller site if the intended value is already in the corresponding register. It would result in incorrect identification of the number of arguments and/or argument widths.

In addition, existing CFI approaches [1, 10, 18] use memory page protection mechanism, Data Execution Prevention (DEP) as an underlying basis. Therefore, they can use read-only tables to store valid targets of indirect branches [10] and insert read-only tags inside the code segment [1]. At runtime, these tables and tags will be checked to see whether the execution follows the policy. However, there are scenarios in which such page-level protection is unavailable, e.g., bare-metal systems which do not have a Memory Management Unit (MMU) and applications with dynamically generated code. Moreover, data race attacks [20], Rowhammer attacks [21], and Data-oriented programming (DOP) [22] have demonstrated that it is possible to gain arbitrary memory read and write access.

Furthermore, CFI-protected programs require extra execution time and space compared to their native counterparts [1, 10, 11]. For example, whenever there is an indirect control-flow transfer, the CFI checks are executed no matter whether there are attacks, thus the protected programs are in general slower than the native versions. For instance, classic CFI reports about 20% performance overhead which hinders its wide adoption.

In this book, we propose approaches to well support these three features. We first systematically study the practicality of recovering fine-grained CFI policies for binaries compiled with different optimization levels. Next, we propose C^3, a novel CFI approach which encodes CFI policies into the machine instructions directly without relying on the assumption that read-only data and code cannot be overwritten. Last but not least, we propose *DynCFI* to enforce CFI based on the dynamic code optimization platform DynamoRIO [23] to improve the performance. The details of these works are presented as follows.

1.2 Practicality of Recovering Fine-Grained CFI Policies

Since having an accurate CFI policy (CFG) is the prerequisite to CFI enforcement, in this book, we do a systematic study on the practicality of recovering fine-grained CFI policies at the binary level. Specifically, we study how compiler optimization would impact the accuracy of CFG construction (function signature recovery) on x86-64 platform. Specifically, we first theoretically analyze the possible ways in which compiler optimizations could impact the accuracy of two most recent approaches in function signature recovery for CFI, namely TypeArmor [18] and τCFI [19], and then experiment with a large number of testing binaries to evaluate the extent to which such complications arise on real-world applications. All testing binaries are obtained by using two commonly used compilers: gcc-8 and clang-7, with different optimization levels ranging from O0 to O3 for x86-64.

The result shows that compiler optimizations have both positive and negative impacts on function signature recovery. For example, optimizations make the identification of variadic functions more accurate. However, compile optimizations could make identification of the number of arguments and the type inferencing at callees less accurate, because of the elimination of unused arguments and promotion/demotion of argument types. In order to mitigate these inaccuracies, we propose our improved policies to recover the function signatures more accurately.

1.3 Control-Flow Carrying Code

A novel CFI approach called C^3 which encodes CFI policies into the machine instructions directly without relying on the assumption that read-only data and code cannot be overwritten is implemented. In it, each basic block in the program is encrypted with a key derived from the CFG. More specifically, the key is derived from the addresses of valid callers of the basic block to ensure correct control-flow transfers. At runtime, only the valid callers (their addresses) could enable the correct reconstruction of the key to decrypt the basic block. The challenge is a basic block may have multiple valid callers, while the successor block has to be encrypted with a single key. In order to enable the reconstruction of the single correct key by all the valid control-flow transfers, secret sharing scheme [24] is used to make the key shared among valid callers.

We apply C^3 to a number of server and non-server applications on the Linux platform. Our experimental results demonstrate that C^3 effectively defends against control-flow hijacking attacks and at the same time, introduces realistic runtime performance overhead for server applications comparable to existing Instruction-Set Randomization (ISR) implementations on the same instrumentation platform.

1.4   Control-Flow Integrity Enforcement Based on Dynamic Code Optimization

A framework that can efficiently enforce CFI based on dynamic code optimization platform is implemented. A lot of well established and mature dynamic code optimizers are proven to introduce minimal overhead, and we believe that they could result in a system that significantly outperforms existing CFI implementation. We enforce a set of security policies on top of DynamoRIO [23] for CFI properties. The results show that it can achieve better performance compared with previous CFI approaches. Moreover, we further investigate the exact contribution to this performance improvement. Specifically, we propose a three-dimensional design space and perform comprehensive experiments to evaluate the contribution of each axis in the design space of performance overhead. The results show that traces in the dynamic optimizer had contributed the most performance improvement. This is because the trace mechanism can avoid some indirect branch lookups by inlining a popular target of an indirect branch into a trace.

1.5   Organization

The reminder of this book is organized as follows: Chap. 2 is a literature review which examines closely related research. Chapter 3 presents details on the systematic evaluation on the extent to which compiler optimization could impact the accuracy of existing approaches in function signature recovery. Chapter 4 describes the system C^3 that embeds the CFI policy into machine instructions. Chapter 5 introduces the framework *DynCFI* that enforce CFI with a dynamic code optimization platform. Chapter 6 summarizes the contribution of this book.

References

1. M. Abadi, M. Budiu, U. Erlingsson, J. Ligatti, Control-flow integrity, in *Proceedings of the 12th ACM Conference on Computer and Communications Security*, (ACM, 2005) pp. 340–353
2. S. Andersen, V. Abella, Data execution prevention. *Changes to functionality in microsoft windows xp service pack*, 2 (2004)
3. P. Team, Pax address space layout randomization. http://pax.grsecurity.net/docs/aslr.txt (2003)
4. C. Cowan, C. Pu, D. Maier, J. Walpole, P. Bakke, S. Beattie, A. Grier, P. Wagle, Q. Zhang, H. Hinton, Stackguard: automatic adaptive detection and prevention of buffer-overflow attacks, in *Proceedings of the 7th USENIX Security Symposium, San Antonio, TX*, vol. 98, pp. 63–78 (1998)
5. J. Pincus, B. Baker, Beyond stack smashing: recent advances in exploiting buffer overruns. *Proceedings of the 25th IEEE Symposium on Security and Privacy*, vol. 2, no. 4, pp. 20–27 (2004)
6. G.F. Roglia, L. Martignoni, R. Paleari, D. Bruschi, Surgically returning to randomized lib (c), in *Proceedings of the 25th Annual Computer Security Applications Conference*, (IEEE, 2009) pp. 60–69
7. T. Bletsch, X. Jiang, V.W. Freeh, Z. Liang, Jump-oriented programming: a new class of code-reuse attack, in *Proceedings of the 6th ACM Symposium on Information, Computer and Communications Security*, (ACM, 2011) pp. 30–40
8. S. Checkoway, L. Davi, A. Dmitrienko, A.-R. Sadeghi, H. Shacham, M. Winandy. Return-oriented programming without returns, in *Proceedings of the 17th ACM Conference on Computer and Communications Security*, (ACM, 2010) pp. 559–572
9. H. Shacham. The geometry of innocent flesh on the bone: return-into-libc without function calls (on the x86), in *Proceedings of the 14th ACM Conference on Computer and Communications Security*, (ACM, 2007) pp. 552–561
10. M. Zhang, R. Sekar, Control flow integrity for cots binaries, in *Proceedings of the 22nd USENIX Security Symposium*, pp. 337–352 (2013)
11. C. Zhang, T. Wei, Z. Chen, L. Duan, L. Szekeres, S. McCamant, D. Song, W. Zou, Practical control flow integrity and randomization for binary executables. In *Proceedings of the 34th IEEE Symposium on Security and Privacy*, (IEEE, 2013) pp. 559–573
12. B. Niu, G. Tan, Modular control-flow integrity. in *Proceedings of the 21st ACM Conference on Computer and Communications Security*, (ACM, 2014) pp. 577–587
13. B. Niu, G. Tan, Per-input control-flow integrity, in *Proceedings of the 22nd ACM Conference on Computer and Communications Security*, (ACM, 2015) pp. 914–926
14. L. Davi, A.-R. Sadeghi, D. Lehmann, F. Monrose. Stitching the gadgets: on the ineffectiveness of coarse-grained control-flow integrity protection, in *Proceedings of the 23rd USENIX Security Symposium* (2014)
15. E. Göktas, E. Athanasopoulos, H. Bos, G. Portokalidis. Out of control: overcoming control-flow integrity, in *Proceedings of the 35th IEEE Symposium on Security and Privacy*, (IEEE, 2014) pp. 575–589
16. F. Schuster, T. Tendyck, C. Liebchen, L. Davi, A.-R. Sadeghi, T. Holz, Counterfeit object-oriented programming: on the difficulty of preventing code reuse attacks in c++ applications, in *Proceedings of the 36th IEEE Symposium on Security and Privacy*, (IEEE, 2015) pp. 745–762
17. C. Tice, T. Roeder, P. Collingbourne, S. Checkoway, Ú. Erlingsson, L. Lozano, G. Pike, Enforcing forward-edge control-flow integrity in {GCC} & {LLVM}, in *Proceedings of the 23rd USENIX Security Symposium*, pp. 941–955 (2014)
18. V. Van Der Veen, E. Göktas, M. Contag, A. Pawoloski, X. Chen, S. Rawat, H. Bos, T. Holz, E. Athanasopoulos, C. Giuffrida. A tough call: mitigating advanced code-reuse attacks at the binary level, in *Proceedings of the 37th IEEE Symposium on Security and Privacy*, (IEEE, 2016) pp. 934–953

19. P. Muntean, M. Fischer, G. Tan, Z. Lin, J. Grossklags, C. Eckert, τ cfi: type-assisted control flow integrity for x86-64 binaries. In *Proceedings of the 21st International Symposium on Research in Attacks, Intrusions, and Defenses*, (Springer, 2018) pp. 423–444
20. M. Zhang, R. Sekar, Control flow and code integrity for cots binaries: an effective defense against real-world ROP attacks, in *Proceedings of the 31st Annual Computer Security Applications Conference*, (Springer, 2018) pp. 91–100
21. E. Bosman, K. Razavi, H. Bos, C. Giuffrida. Dedup est machina: memory deduplication as an advanced exploitation vector, in *Proceedings of the 37th IEEE Symposium on Security and Privacy*, (IEEE, 2016) pp. 987–1004
22. H. Hu, S. Shinde, S. Adrian, Z.L. Chua, P. Saxena, Z. Liang, Data-oriented programming: on the expressiveness of non-control data attacks, in *Proceedings of the 37th IEEE Symposium on Security and Privacy*, (IEEE, 2016) pp. 969–986
23. D. Bruening, *Efficient,Transparent,and Comprehensive Runtime Code Manipulation*. Ph.D. thesis, Massachusetts Institute of Technology (2004)
24. A. Shamir, How to share a secret. Commun. ACM **22**(11), 612–613 (1979)

Chapter 2
Literature Review

2.1 Control-Flow Hijacking

C and C++ are perhaps the most important programming languages due to its high performance. However, these programs all suffer from memory corruption issues such as out-of-bound accesses or object use-after-free bugs.

Listing 2.1 shows a real memory corruption bug found in `MiniUPnP` 1.0. An attacker has full control over `action` at line 6. The code at line 10 copies `methodlen` size of data from the address pointed by p to the address pointed by `method`. Hence, if the attacker passes a long quoted method (more than 2048 bytes), it will cause buffer overflow, which results in code-injection and code-reuse attacks.

```
1  ExecuteSoapAction( struct upnphttp *h, const char *action,, int n)
2  {
3      char *p;
4      char method[2048];
5      ...
6      p = strchr(action,'#');
7      methodlen = strchr(p,'"') - p - 1;
8      ...
9      memset(method, 0, 2048);
10     memcpy(method, p, methodlen);
11     ...
12 }
```

Listing 2.1 A stack buffer overflow bug (CVE-2013-0230) in MiniUPnP 1.0

Code-Injection Attack. In code injection attacks, the attacker injects new code into the address space of the victim program and executes her code. In the example of Listing 2.1, if the stack is executable, the attacker could inject the shellcode to the `method` array and the return address of the current function can also be overwritten with a pointer pointing to the entry of the injected code. Then, when the current function returns, instead of returning to the caller, the function returns to the injected code and executes it. This kind of control-flow hijacking attack has been mitigated by

© The Author(s), under exclusive license to Springer Nature Switzerland AG 2021
Y. Lin, *Novel Techniques in Recovering, Embedding, and Enforcing Policies for Control-Flow Integrity*, Information Security and Cryptography, https://doi.org/10.1007/978-3-030-73141-0_2

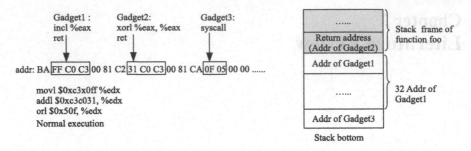

Fig. 2.1 A proof of concept example of Return-Oriented Programming attacks

some effective mechanisms, such as Data Execution Prevention [1] and Instruction-set randomization [2–4].

Code-Reuse Attack. In code-reuse attacks, rather than directly injecting shellcode to the victim, the attacker chains the existing code bytes (gadgets) together to perform the malicious operation, such as bypassing DEP, so that she can execute the injected shellcode. In the example in Listing 2.1, the attacker could simply overwrite the return address to the address of a libc function system and write arguments to the function on the stack. When the function returns, system will be executed with attacker-fed arguments, which enable the attacker to execute arbitrary commands.

Usually, simply overwriting a return address is not sufficient to mount an attack. Attackers use a more advanced code reuse technique called Return-Oriented Programming (ROP) [5] to perform any malicious operation. To mount such an attack, the attackers first scan the code bytes and find gadgets, which are instruction sequences ending with a return (or an indirect call/jump), and perform basic operations such as an addition or memory load. Then, by carefully overflowing the stack (heap), the attackers can chain these gadgets into an arbitrary program.

We use a simple proof-of-concept example in Fig. 2.1 to demonstrate principles of ROP attacks in x86-64 Linux. Suppose function foo has a stack overflow vulnerability as shown in Listing 2.1, so the attacker can overwrite the stack buffer. As shown, there is a sequence of instructions which perform some basic arithmetic operations at address addr. However, if we decode the instruction from the middle, three ROP gadgets will be found. Gadget1 increments register %eax by one and returns; Gadget2 sets %eax to zero; and Gadget3 performs a system call. If the attacker overwrite the return address of function foo to the address of Gadget2, and write 32 copies of Gadget1 addresses, followed by the address of Gadget3, the attackers can essentially set %eax to 0, increment it by 32, and execute a system call after foo returns. Since Linux uses %eax to pass the system call number to the kernel, and 32 is the system call number for pause, the current process will sleep.

A lot of mitigation techniques are deployed to defend against this kind of attack, including stack canaries and Address Space Layout Randomization (ASLR).

2.2 Deployed Defenses

Stack Canaries. Stack canaries [6] mitigate control-flow hijacking attacks by monitoring the integrity of return addresses. It inserts a canary before every return address and checks the value of the canary before a function returns. If the canary changes, stack overflow might have happened and the program is terminated. However, stack canary can be bypassed. As shown in BROP [7], since the cookie is a randomly chosen value, it may be quickly guessed. Meanwhile, it cannot protect heap buffer overflows.

Data Execution Prevention (DEP). DEP [1] prevents code-injection attacks by enforcing the stack/heap memory is non-executable, so the attacker cannot directly write shellcode to these memory pages. We can find DEP has no protection against code reuse attacks. Moreover, DEP is not compatible with programs that generate and modify code on-the-fly, such as Just-In-Time (JIT) compilers.

Address Space Layout Randomization (ASLR). The basic idea of ASLR [8] is to make it harder for attackers to precisely locate the reusable code. Specifically, program modules such as the executable file and dependent libraries are compiled to be position-independent and loaded into randomized addresses. However, ASLR can be bypassed by information disclosure and brute force attacks [9]. Some fine-grained code layout randomization approaches [10–12] are proposed, in which permutation on functions, instruction layout, basic blocks, and code transformation are implemented. However, they are vulnerable to JIT-ROP [13].

2.3 Control-Flow Integrity

Stronger than all the deployed defenses, Control-Flow Integrity (CFI) forces control-flow transfers in the program to follow the policy presented by the CFG.

Control-flow transfers can be either direct or indirect. Direct edges include sequential instruction execution and direct branching. For example, the transfer from a direct call instruction to its target function address is a direct control-flow transfer. Since targets of direct control transfers cannot be arbitrarily controlled by attackers, they are less of concern. Indirect transfers through indirect branch instructions including indirect calls, indirect jumps, and returns are more dangerous, because their targets may be arbitrarily controlled by attackers. To ensure CFI for indirect branches, they are checked before execution so that their targets are always legal.

In general, it can be classified into two categories: coarse-grained CFI and fine-grained ones.

2.3.1   Coarse-Grained CFI

Having accurate and practical enforcement of CFI is known to be hard. First, it is generally difficult to accurately identify the target locations for all control transfers. Existing solutions typically apply a coarse-grained policy (e.g., to allow indirect calls to any functions) for Commercial Off-The-Shelf (COTS) software whose source code is unavailable. This kind of CFI marks the valid targets of indirect control transfers with unique identifiers (IDs) and then inserts ID-checks into the program before each indirect branch transfer. An indirect branch is allowed to jump to any destination with the same ID.

CFIMon [14] uses three IDs for all indirect branch transfers. The target of a return instruction can be any call-preceded basic blocks and the target of an indirect call can be any function. The valid targets for indirect jumps are obtained by making use of online training. It leverages Branch Trace Store (BTS) mechanism [15] supported by hardware to collect in-flight control transfers, and once the BTS buffer is nearly full, a monitor process will start to compare them with the valid targets to decide whether there exists an attack.

BinCFI [16] uses two IDs for all indirect branch transfers: one for ret and indirect jump instructions, another for indirect call instructions. All indirect branches are instrumented to jump to the corresponding address translation routine that determines the targets of the transfers, if one target cannot be found, it means there is a control-flow hijacking attack. However, whenever there is a control-flow transfer, the CFI checks are always executed. Our proposed approach *DynCFI* does not require to perform CFI check for each control-flow transfer due to the trace mechanism used in the dynamic code optimization platform DynamoRIO [17]. Specifically, the trace mechanism can avoid some indirect branch lookups by inlining a popular target of an indirect branch into a trace.

CCFIR [18] implements a 3-IDs approach, which extended the 2-IDs approach by further separating returns to sensitive and non-sensitive functions. All control-flow targets for indirect branches are collected and randomly allocated on a so-called springboard section, and indirect branches are only allowed to use control-flow targets contained in the springboard section. CCFIR can manage the indirect branch transfers better, and the targets of indirect branches are more restricted than other approaches. However, in the springboard section, there are other indirect branches, and memory disclosure can reveal the content of the entire springboard section, which can be leveraged by attackers.

BinCC [19] enforces a 4-IDs approach by dividing the binary code into several mutually exclusive code continents and further classifying each indirect transfer within a code continent as either an Intra-Continent transfer or an Inter-Continent transfer. Different continent transfers will have different valid targets. For example, the valid targets of Intra-Continent transfer are always inside their own code continent. For other Inter-Continent transfers, the policy is indirect call nodes can only reach all root nodes in all continents, indirect return nodes can only reach indirect call nodes, and indirect jump nodes can only reach all root and call nodes.

Since this kind of CFI approaches recover a coarse-grained CFG, they may be bypassed by some advanced code-reuse attacks [20–22]. A lot of fine-grained CFI approaches are proposed.

2.3.2 Fine-Grained CFI

Most of these CFI approaches rely on the availability of source code. MCFI [23] propagates source-level information such as type information to the binary level as metadata, and gathers such metadata at program load time to build a precise CFG, which is consulted (or read) by the program to detect CFI violations. Specifically, when a code module is loaded during execution, the loading module's metadata is combined with the loaded module's metadata to compute a CFG for both modules, and then the old CFG will be replaced with the new CFG. πCFI [24] starts a program with an empty CFG and let the program itself lazily compute the CFG on the fly. The main idea behind this empty CFG approach is to affix edges on runtime prior to being used for branch instructions.

Forwarding CFI [25] protects binaries by inserting checks before all forward-edge control-flow transfers to check whether the function signatures (return type and the number of arguments) are correct. Cryptographically enforced CFI [26] enforces another form of fine-grained CFI by adding a message authentication code (MAC) that is computed with type information to control-flow elements, which prevents the usage of unintended control-flow transfers in the CFG.

Since the requirement of source code makes these approaches difficult to be deployed. CFI based on function signature matching at the binary level is proposed. There is two work focus on it called TypeArmor [27] and τCFI [28]. TypeArmor is the first work that uses the function signature on the number of arguments at the binary level to enforce CFI. It extracts the number of arguments both at the caller and callee sites by performing backward and forward analysis, and the target of the indirect call can only be functions that have matching signatures. τCFI is the follower of TypeArmor that tries to construct a more fine-grained CFG by combining the width of the argument registers as the additional function signature.

Both of them rely on the x86-64 calling convention that the first six arguments for integer are passed through registers (assuming System V ABI). However, the compiler may generate code that violates the calling convention. For example, modern compilers may not set the argument register explicitly at the caller site if it finds the argument value is already in the corresponding register. We systematically study how compiler would impact function signature recovery for TypeArmor [27] and τCFI [28] and find compiler optimizations have a great impact on function signature recovery. Such errors could lead to invalid function calls being allowed or, even worse, valid calls being inadvertently blocked.

Both coarse-grained and fine-grained approaches usually have high performance overhead as for every indirect control-flow transfer they need to do CFI check. Moreover, they need to add CFI checks into the code section or consult read-only data structures. If these CFI checks and read-only data structures are compromised, these mitigation approaches can be bypassed easily. Our proposed system C^3 embeds the CFI policy into every machine instruction without relying on the assumption of keeping such meta read-only. Furthermore, all of them do not insert CFI checks for unintended control-flow transfers, making them being bypassed by the exploit proposed by Conti et al. [29]. Such exploits would not work on C^3 as all instructions (intended or unintended) are encrypted.

2.4 Instruction-Set Randomization

Instruction-Set Randomization (ISR) was initially proposed to fight against code-injection attacks [2–4, 30]. It encrypts instructions and provides a unique instruction set to every program. Injected code would first be decrypted to a random byte sequence and result in illegal instructions executed. Recently, researchers look into using ISR to defend against code-reuse attacks (CRA). Scylla [31] encrypts every instruction in a basic block with respect to its predecessor to defend against CRA that jumps to the middle of a basic block. However, it does not stop attacks that make use of the entire basic block to construct gadgets. Polyglot [32] combines ISR with fine-grained code randomization to defend against JIT-ROP [13]. It encrypts every basic block of instructions by XORing them with the starting address of the block. Since the encryption key is only derived from the address of the basic block and not tied to control transfers, CFI can be compromised with control transfers from invalid callers. C^3, on the other hand, is designed for enforcing CFI with encryption key tied to all valid control transfers. Invalid callers will result in the wrong decryption key generated and random code bytes obtained. SOFIA [33] uses ISR to enforce CFI for cyber-physical systems with instructions at a fixed length of 32-bit via an integrity check of instruction blocks where the Message Authentication Code (MAC) is encrypted. SOFIA requires access to source code of the program to be protected in order to rearrange the CFG so that every basic block has up to two callers. On the other hand, we propose a novel idea C^3 which uses secret sharing to support multiple (potentially more than two) callers.

2.5 Function Signature Recovery

Besides TypeArmor [27], liveness analysis and heuristic methods based on calling conventions and idioms were used to recover function signatures. ElWazeer et al. [34] apply liveness analysis to recover arguments, variables, and their types for x86 executables. It assumes all registers are arguments to every function and then traverse the call graph of the executable in post-order depth-first search traversal to check whether there is a "real" use of this register in the function. Only "real" used registers are considered as arguments. The argument that stored on the stack is recognized by making use of Value Set Analysis [35]. Points-to analysis is used to recover the type of an argument (variable) according to some type recovering rules.

TIE [36] infers variable types in binaries through formulating the usage of different data types. It first lifts the binary code to a binary analysis language called BIL by using BAP [37], and then type information is inferred by solving type constraints. It depends on that some known prototype functions will be called (e.g., library functions), so that the constraint can be solved by making use of some rules. Caballero et al. [38] make use of dynamic liveness analysis to recover function arguments for execution traces. Since it is a dynamic analysis, it cannot guarantee the full coverage of unused arguments during an execution trace. Recently, Zeng et al. [39] propose to perform type inference based on debugging information generated by the compiler so that a high-precision CFG can be constructed to help CFI enforcement. Another direction is to make use of machine learning approaches to recover function signatures. For example, EKLAVYA [40] uses a three layers Recurrent Neural Network to learn the number and types of arguments from disassembled binary code.

2.6 Non-control Data Attacks

Compared to control-flow hijacking attacks, Non-control attacks manipulate non-control data to alter a programs benign behavior without violating its control-flow integrity. This kind of non-control data attacks can cause significant damages, such as leakage of secret keys (HeartBleed) [41] and enabling untrusted code import [42]. Hu et al. [43] make use of data-flow stitching to systematically finds ways to join data flows in the program to generate data-oriented exploits. Specifically, it takes as input a vulnerable program with a memory error, an input that exploits that memory error, and a benign input that triggers the same execution path (two-dimensional data-flow graph (2D-DFG)), and uses backward and forward slicing to pinpoint data flow paths between inputs and pre-identified sensitive data.

In a follow up work, Hu et al. [44] propose data-oriented programming (DOP) which demonstrates that non-control data attacks resulting from a single memory error can be Turing-complete, and a large number of data-oriented gadgets can be found. These gadgets require to deliver operation result with memory and must execute in at least one legitimate control flow, and need not execute immediately one after another. By stitching these gadgets together using a dispatcher (e.g., loop), the attacker can perform arbitrary operations. It has shown that DOP can be used to change the permissions of read-only pages to bypass current CFI implementations by triggering dlopen's internal gadgets to corrupt the read-only pages.

References

1. S. Andersen, V. Abella, Data execution prevention, in *Changes to Functionality in Microsoft Windows XP Service Pack*, vol. 2 (2004)
2. G.S. Kc, A.D. Keromytis, V. Prevelakis, Countering code-injection attacks with instruction-set randomization, in *Proceedings of the 10th ACM Conference on Computer and Communications Security* (ACM, 2003), pp. 272–280
3. E.G. Barrantes, D.H. Ackley, T.S. Palmer, D. Stefanovic, D.D. Zovi, Randomized instruction set emulation to disrupt binary code injection attacks, in *Proceedings of the 10th ACM Conference on Computer and Communications Security* (ACM, 2003), pp. 281–289
4. G. Portokalidis, A.D. Keromytis, Fast and practical instruction-set randomization for commodity systems, in *Proceedings of the 26th Annual Computer Security Applications Conference* (ACM, 2010), pp. 41–48
5. H. Shacham, The geometry of innocent flesh on the bone: return-into-libc without function calls (on the x86), in *Proceedings of the 14th ACM Conference on Computer and Communications Security* (ACM, 2007), pp. 552–561
6. C. Cowan, C. Pu, D. Maier, J. Walpole, P. Bakke, S. Beattie, A. Grier, P. Wagle, Q. Zhang, H. Hinton, StackGuard: automatic adaptive detection and prevention of buffer-overflow attacks, in *Proceedings of the 7th USENIX Security Symposium*, vol. 98 (San Antonio, TX, 1998), pp. 63–78
7. A. Bittau, A. Belay, A. Mashtizadeh, D. Mazières, D. Boneh, Hacking blind, in *Proceedings of the 35th IEEE Symposium on Security and Privacy* (IEEE, 2014), pp. 227–242
8. P. Team, Pax address space layout randomization (2003), http://pax.grsecurity.net/docs/aslr.txt
9. H. Shacham, M. Page, B. Pfaff, E.-J. Goh, N. Modadugu, D. Boneh, On the effectiveness of address-space randomization, in *Proceedings of the 11th ACM Conference on Computer and Communications Security* (ACM, 2004), pp. 298–307
10. J. Hiser, A. Nguyen-Tuong, M. Co, M. Hall, J.W. Davidson, ILR: where'd my gadgets go? in *Proceedings of the 33th IEEE Symposium on Security and Privacy* (IEEE, 2012), pp. 571–585
11. V. Pappas, M. Polychronakis, A.D. Keromytis, Smashing the gadgets: hindering return-oriented programming using in-place code randomization, in *Proceedings of the 33th IEEE Symposium on Security and Privacy* (IEEE, 2012), pp. 601–615
12. R. Wartell, V. Mohan, K.W. Hamlen, Z. Lin, Binary stirring: self-randomizing instruction addresses of legacy x86 binary code, in *Proceedings of the 19th ACM conference on Computer and communications security* (ACM, 2012), pp. 157–168
13. K.Z. Snow, F. Monrose, L. Davi, A. Dmitrienko, C. Liebchen, A.-R. Sadeghi, Just-in-time code reuse: on the effectiveness of fine-grained address space layout randomization, in *Proceedings of the 34th IEEE Symposium on Security and Privacy* (IEEE, 2013), pp. 574–588
14. Y. Xia, Y. Liu, H. Chen, B. Zang, CFIMon: detecting violation of control flow integrity using performance counters, in *Proceedings of the 42nd Annual IEEE/IFIP International Conference on Dependable Systems and Networks* (IEEE, 2012), pp. 1–12
15. P. Guide, Intel® 64 and IA-32 architectures software developer's manual (2016)
16. M. Zhang, R. Sekar, Control flow integrity for cots binaries, in *Proceedings of the 22nd USENIX Security Symposium* (2013), pp. 337–352
17. D. Bruening, *Efficient,Transparent,and Comprehensive Runtime Code Manipulation*. PhD thesis, Massachusetts Institute of Technology (2004)
18. C. Zhang, T. Wei, Z. Chen, L. Duan, L. Szekeres, S. McCamant, D. Song, W. Zou, Practical control flow integrity and randomization for binary executables, in *Proceedings of the 34th IEEE Symposium on Security and Privacy* (IEEE, 2013), pp. 559–573
19. M. Wang, H. Yin, A.V. Bhaskar, P. Su, D. Feng, Binary code continent: finer-grained control flow integrity for stripped binaries, in *Proceedings of the 31st Annual Computer Security Applications Conference* (ACM, 2015), pp. 331–340
20. N. Carlini, D. Wagner, ROP is still dangerous: breaking modern defenses, in *Proceedings of the 23rd USENIX Security Symposium* (2014), pp. 385–399

21. E. Göktas, E. Athanasopoulos, H. Bos, G. Portokalidis, Out of control: overcoming control-flow integrity, in *Proceedings of the 35th IEEE Symposium on Security and Privacy* (IEEE, 2014), pp. 575–589
22. F. Schuster, T. Tendyck, J. Pewny, A. Maaß, M. Steegmanns, M. Contag, T. Holz, Evaluating the effectiveness of current anti-ROP defenses, in *Proceedings of the 17th International Workshop on Recent Advances in Intrusion Detection* (Springer, 2014), pp. 88–108
23. B. Niu, G. Tan, Modular control-flow integrity, in *Proceedings of the 21st ACM Conference on Computer and Communications Security* (ACM, 2014), pp. 577–587
24. B. Niu, G. Tan, Per-input control-flow integrity, in *Proceedings of the 22nd ACM Conference on Computer and Communications Security* (ACM 2015), pp. 914–926
25. C. Tice, T. Roeder, P. Collingbourne, S. Checkoway, Ú. Erlingsson, L. Lozano, G. Pike, Enforcing forward-edge control-flow integrity in {GCC} & {LLVM}, in *Proceedings of the 23rd USENIX Security Symposium* (2014), pp. 941–955
26. A.J. Mashtizadeh, A. Bittau, D. Boneh, D. Mazières, CCFI: cryptographically enforced control flow integrity, in *Proceedings of the 22nd ACM Conference on Computer and Communications Security* (ACM, 2015), pp. 941–951
27. V. Van Der Veen, E. Göktas, M. Contag, A. Pawoloski, X. Chen, S. Rawat, H. Bos, T. Holz, E. Athanasopoulos, C. Giuffrida, A tough call: mitigating advanced code-reuse attacks at the binary level, in *Proceedings of the 37th IEEE Symposium on Security and Privacy* (IEEE, 2016), pp. 934–953
28. P. Muntean, M. Fischer, G. Tan, Z. Lin, J. Grossklags, C. Eckert, τ CFI: type-assisted control flow integrity for x86-64 binaries, in *Proceedings of the 21st International Symposium on Research in Attacks, Intrusions, and Defenses* (Springer, 2018), pp. 423–444
29. M. Conti, S. Crane, L. Davi, M. Franz, P. Larsen, M. Negro, C. Liebchen, M. Qunaibit, A.-R. Sadeghi, Losing control: on the effectiveness of control-flow integrity under stack attacks, in *Proceedings of the 22nd ACM Conference on Computer and Communications Security* (ACM, 2015), pp. 952–963
30. J. Fu, X. Zhang, Y. Lin, An instruction-set randomization using length-preserving permutation, in *Proceedings of the 14th IEEE International Conference on Trust, Security and Privacy in Computing and Communications* (IEEE, 2015), pp. 376–383
31. D. Sullivan, O. Arias, D. Gens, L. Davi, A.-R. Sadeghi, Y. Jin, Execution integrity with in-place encryption. *arXiv preprint* arXiv:1703.02698 (2017)
32. K. Sinha, V. P. Kemerlis, S. Sethumadhavan, Reviving instruction set randomization, in *Proceedings of the 10th International Symposium on Hardware Oriented Security and Trust* (IEEE, 2017), pp. 21–28
33. R. de Clercq, J. Götzfried, D. Übler, P. Maene, I. Verbauwhede, Sofia: software and control flow integrity architecture. Comput. Secur. **68**, 16–35 (2017)
34. K. ElWazeer, K. Anand, A. Kotha, M. Smithson, R. Barua, Scalable variable and data type detection in a binary rewriter, in *ACM SIGPLAN Conference on Programming Language Design and Implementation* (2013), pp. 51–60
35. G. Balakrishnan, T. Reps, Analyzing memory accesses in x86 executables, in *Proceedings of the 13rd International Conference on Compiler Construction* (Springer, 2004), pp. 5–23
36. J. Lee, T. Avgerinos, D. Brumley, Tie: principled reverse engineering of types in binary programs, in *Proceedings of the 18th Network and Distributed System Security Symposium* (2011)
37. D. Brumley, I. Jager, T. Avgerinos, E.J. Schwartz, Bap: a binary analysis platform, in *Proceedings of the 23rd International Conference on Computer Aided Verification* (Springer, 2011), pp. 463–469
38. J. Caballero, N.M. Johnson, S. McCamant, D. Song, Binary code extraction and interface identification for security applications. Technical report, California Univ Berkeley Dept of Electrical Engineering and Computer Science (2009)
39. D. Zeng, G. Tan, From debugging-information based binary-level type inference to CFG generation, in *Proceedings of the 8th ACM Conference on Data and Application Security and Privacy* (ACM, 2018), pp. 366–376

40. Z.L. Chua, S. Shen, P. Saxena, Z. Liang, Neural nets can learn function type signatures from binaries, in *Proceedings of the 26th USENIX Security Symposium* (2017), pp. 99–116
41. The heartbleed bug. http://heartbleed.com/
42. Y. Yang, ROPS are for the 99 (2014), https://cansecwest.com/slides/2014/ROPsareforthe99 CanSecWest2014.pdf
43. H. Hu, Z. L. Chua, S. Adrian, P. Saxena, Z. Liang, Automatic generation of data-oriented exploits, in *Proceedings of the 24th USENIX Security Symposium* (2015), pp. 177–192
44. H. Hu, S. Shinde, S. Adrian, Z.L. Chua, P. Saxena, Z. Liang, Data-oriented programming: on the expressiveness of non-control data attacks, in *Proceedings of the 37th IEEE Symposium on Security and Privacy* (IEEE, 2016), pp. 969–986

Chapter 3
When Function Signature Recovery Meets Compiler Optimization

3.1 Introduction

Control-Flow Integrity (CFI) [1] is a promising technique in defending against control-flow hijacking attacks [2–5] by enforcing that runtime control flows follow valid paths in the program's Control-Flow Graph (CFG). Many approaches [6–9] opt for fine-grained CFGs obtained at compilation time due to their high accuracy. However, it is difficult to precisely recover CFGs at the binary level since compilers do not preserve much information in the process of compilation [10]. Most existing approaches had to conservatively consider all functions as potential targets of an indirect caller, resulting in loosened CFI policies [11, 12] which make these approaches vulnerable to various attacks [13–17].

Latest approaches [18, 19] recover function signatures at the binary level by following calling conventions and only allow control flows between callees and callers with matching function signatures. Although generally good accuracy had been reported, e.g., TypeArmor [19] achieved 83.26% and 79.19% accuracy in identifying the number of arguments at callees and callers, respectively, in this paper, we challenge this belief of high accuracy when dealing with *optimized* binary executables. We subject TypeArmor to the same set of applications as chosen in the original paper, which are now compiled with different compiler versions with new optimization strategies enabled and find that the accuracy drops to 72.89% and 72.27%. The accuracy goes even lower to 63.74% and 69.36% when analyzing more complicated applications (e.g., Binutils) even with the same compiler version used in the original paper.

Our further investigation shows that this is because compiler optimizations may violate calling conventions and result in unmatched function signatures recovered at valid callees and callers. For example, modern compilers may not set or reset an argument register explicitly at the caller if the intended value is already in the corresponding register. The non-existence of the value assignment instruction therefore confuses the recovery process and results in underestimation on the number of function arguments. As shown in Listing 3.1, the indirect call at line 2 has 4 arguments, but the compiled binary code (with optimization flag -O2 by clang) does

```
 1  long test(long a, long b, long c, long d, long e, long f) {
 2      long sum1 = (*fptr1)(a,b,c,d);
 3      ......
 4      //function ldiv returns a struct
 5      ldiv_t ldivrs;
 6      rs = ldiv (1000000L,132L);
 7      long sum2 = (*fptr2)(a,rs.quot, rs.rem);
 8      if (sum2 > sum1)
 9          return sum2;
10      else
11          return sum1;
12  }
13  0000000000400650 <test>:
14  ......
15  40065b: mov     %r9,(%rsp)
16  40065f: mov     %r8,%r12
17  400662: mov     %rcx,%r13
18  400665: mov     %rdx,%rbp
19  400668: mov     %esi,%r15d
20  40066b: mov     %rdi,%r14
21  40066e: callq   *0x200e04(%rip) # 601478 <fptr1>
22  ......
23  40069e: mov     %r14d,%edi
24  4006a1: mov     %eax,%esi
25  4006a3: callq   *0x2009b7(%rip) # 601060 <fptr2>
```

Listing 3.1 An example when function signature recovery meet compiler optimization.

not prepare for any argument as shown at Line 15–20. Similarly, the compiler only sets the first two arguments (`%edi`, `%esi`) for the indirect call at Line 25 while it requires 3 arguments as shown at Line 7. Such errors in function signature recovery could lead to invalid function calls being allowed or, even worse, valid calls being inadvertently blocked.

In this chapter, we systematically study how compiler optimizations impact the accuracy of function signature recovery on x86-64 platform, with obfuscated binary out of our scope since existing work has clearly shown how obfuscated code complicates static binary analysis [20]. Specifically, we first theoretically analyze the possible ways in which compiler optimizations could impact the accuracy of two most recent approaches in function signature recovery for CFI, namely TypeArmor [19] and τCFI [18], and then experiment with a large number of applications including Binutils,[1] LLVM test-suite,[2] as well as C/C++ applications from Github to evaluate the extent to which such complications arise on real-world applications.

---

[1]https://www.gnu.org/software/binutils/.
[2]https://llvm.org/docs/TestSuiteGuide.html.

We recover the ground truth of function signatures of 552 C and 792 C++ applications compiled with `gcc-8` and `clang-7` with optimization levels `-O0` to `-O3` and compare them with results of TypeArmor [19], τCFI [18], and Ghidra [21] in recovering the number of arguments and argument types.

Results show that compiler optimizations have both positive and negative impacts on function signature recovery. First, optimizations make the identification of variadic functions more accurate as arguments are more likely to be moved to callee-saved registers than being moved onto the stack. At the same time, the elimination of redundant instructions due to optimization also simplifies the argument analysis at callers. However, compiler optimization could make identification of the number of arguments and the type inferencing at callees less accurate, because of the elimination of unused arguments and promotion/demotion of argument types.

In order to mitigate these inaccuracies, we propose our improved policies to recover the function signatures more accurately from optimized binaries. We evaluate our proposed policies with the same set of real-world applications and compare our accuracy with that of existing ones. Results show that, e.g., the likelihood of misidentifying variadic functions in C is reduced from 3.3% to 1.2%. Moreover, our policy can mitigate all issues caused by argument type demotion at callers and argument type promotion at callees. Finally, we look at the bigger picture of CFI policies recovered from binary executables and program source, empirically analyze the implication of errors they make, and reveal scenarios in which compiler optimization makes the task of accurate function signature recovery undecidable.

3.2 Background and Unified Notation

On Linux ×86-64, all arguments of a function are passed from the caller to the callee who is assumed to process every argument. Integer arguments are passed in registers `%rdi`, `%rsi`, `%rdx`, `%rcx`, `%r8`, `%r9` in sequence, while `%XMM0` – `%XMM7` are used to pass floating-point arguments [22]. Additional arguments are pushed onto the stack in reverse order. The return value is stored in `%rax` with potentially the higher 64 bits stored in `%rdx`. Floating-point return values are similarly stored in `%XMM0` and `%XMM1`. Both TypeArmor and τCFI adhere to these calling conventions and do not consider deviations from them.

Variadic functions (such as `printf` in the C library) are used to maximize flexibility in argument passing. These functions accept a variable number of arguments which do not necessarily have fixed types.

TypeArmor [19] and τCFI [18] reconstruct both callee and caller signatures by performing static binary analysis and then use this information to enforce Control-Flow Integrity between callees and callers with similar signatures. TypeArmor uses the number of arguments as the signature, while width (number of bits $p \in \{64, 32, 16, 8\}$) of the argument-storing registers is used by τCFI. Just like in existing approaches, we focus on function signature recovery for integer arguments and use $i \in [1, 6]$ to index the six argument registers. Here we introduce our unified notation to describe the CFI policies TypeArmor and τCFI employ as well as our improved policy (see Sect. 3.5).

3.2.1 Analysis of Callees

Analysis of a callee function typically starts from the function entry and continues in a forward manner until the end of the function. Here, the analysis focuses on the *first* instruction involving a parameter-passing register, which could have one of the following four possible states: $s^{EE} \in \{\dot{w}(), r\dot{w}(), r\dot{w}2s(), c\}$ (we use the dot above a state to denote that it's the analysis result of the *first* instruction involving the corresponding register).

Definition 1. *State $\dot{w}_i(p)$ if the first instruction involving register i is writing into the lower p bits of register i.*

Definition 2. *State $r\dot{w}_i(p)$ if the first instruction involving register i is reading the lower p bits of it and writing to anther register or a non-stack address.*

Definition 3. *State $r\dot{w}2s_i(p)$ if the first instruction involving register i is reading the lower p bits of it and writing to a stack address.*

Definition 4. *State c_i if register i is not involved in any instructions.*

Definition 5. *Argument register state vector observed at callee $P_{EE}^{OB} = <s_1^{EE}, s_2^{EE}, s_3^{EE}, s_4^{EE}, s_5^{EE}, s_6^{EE}>$ where $s_i^{EE} \in \{\dot{w}_i(), r\dot{w}_i(), r\dot{w}2s_i(), c_i\}$ for $i \in [1, 6]$.*

Definition 6. *$b2b_i$ is true if $s_i^{EE} = r\dot{w}2s_i()$ and $s_{i+1}^{EE} = r\dot{w}2s_{i+1}()$ and the corresponding instructions involving registers i and i + 1 are back to back.*

3.2.2 Analysis of Callers

Analysis of a caller function starts at the indirect call instruction and continues in a backward manner until it hits another function call instruction. This backward analysis follows the CFG and focuses on *all instructions* involving the parameter-passing register instead of only the first instruction as in the analysis of callees.

Definition 7. *State $w_i(p)$ if there is an instruction writing to the lower p bits of register i.*

Definition 8. *State $\hat{w}_i$ if there is no instruction writing to register i.*

Definition 9. *Argument register state vector observed at caller $P_{ER}^{OB} = <s_1^{ER}, s_2^{ER}, s_3^{ER}, s_4^{ER}, s_5^{ER}, s_6^{ER}>$ where $s_i^{ER} \in \{w_i(), \hat{w}_i\}$ for $i \in [1, 6]$.*

3.2.3 TypeArmor's Policy on the Number of Arguments

At a callee, TypeArmor [19] performs a forward recursive analysis from the entry block to find out states of the six argument registers. If the state of the sixth argument register ($\%r9$) is $r\dot{w}2s_6()$, TypeArmor concludes that this function is variadic and the number of arguments is the maximal i that makes $b2b_i$ false. If the state of $\%r9$ is not $r\dot{w}2s_6()$, the function is considered non-variadic and the number of arguments is the maximal i with state $r\dot{w}2s_i()$ or $r\dot{w}_i()$.

Definition 10. *The observed number of arguments at callee* $|P_{EE}^{OB}|$ *is:*

$$
\begin{cases}
\underset{i}{\mathrm{argmax}}(\neg b2b_i) & \text{if } s_6^{EE} = r\dot{w}2s_6() \\
\max(\underset{i}{\mathrm{argmax}}(r\dot{w}2s_i()), \underset{i}{\mathrm{argmax}}(r\dot{w}_i())) & \text{otherwise}
\end{cases}
$$

TypeArmor iterates over each indirect caller and performs a backward static analysis to detect the number of arguments prepared. If the states of all argument registers are $w()$, TypeArmor stops the analysis and considers that the caller prepares the maximum number of arguments. If some argument registers are neither $w()$ nor $\hat{w}$, TypeArmor performs a recursive backward analysis on incoming control flows. In cases where incoming control flows are via indirect calls and therefore backward analysis fails in identifying the caller function, TypeArmor assumes that the maximum number of arguments is prepared. It also assumes that the argument registers are always reset between two function calls, and therefore analysis is terminated when a return edge is encountered. In summary, the number of arguments at the caller is the minimal i with state $\hat{w}_i$ minus one.

Definition 11. *The observed number of arguments at caller* $|P_{ER}^{OB}|$ *is:*

$$
\begin{cases}
\underset{i}{\mathrm{argmin}}(s_i^{ER} = \hat{w}_i) - 1 & \text{if } \exists \hat{w}_i \in P_{ER}^{OB} \\
6 & \text{otherwise}
\end{cases}
$$

Since there could be overestimation at callers and underestimation at callees, TypeArmor allows caller A to call callee B if and only if $|P_{ER-A}^{OB}| \geq |P_{EE-B}^{OB}|$.

3.2.4 τCFI's Policy on the Width of Arguments

τCFI [18] is the follower of TypeArmor that constructs a more fine-grained CFG by additionally considering the widths of argument registers as function signatures. It analyzes the number of bits of argument registers that are read or written to at callees and callers, respectively. We use $|s^{EE}|$ and $|s^{ER}|$ to represent the with of arguments at callees and callers, respectively. For example, if $P_{EE}^{OB} = <\dot{w}_1(), \dot{w}_2(), \dot{w}_3(), \dot{w}_4(),$ $r\dot{w}_5(64), r\dot{w}_6(64)>$, then $|s_1^{EE}| = |s_2^{EE}| = |s_3^{EE}| = |s_4^{EE}| = 0$ and $|s_5^{EE}| = |s_6^{EE}| = 64$.

Since the analysis could cause overestimation at callers and underestimation at callees, the CFI policy of τCFI is: caller A can transfer control flow to callee B if and only if: $\forall i \in [1, |P_{ER}^{OB}|], |s_i^{ER}| >= |s_i^{EE}|$.

We also denote the ground truth for the states of argument registers at callees and callers as P_{EE}^{GT} and P_{ER}^{GT}, respectively. $|s^{EE,GT}|$ and $|s^{ER,GT}|$ are used to denote the ground truth on the width of arguments.

3.3 Eight Ways in Which Compiler Optimization Impacts Function Signature Recovery

In this section, we present our analysis in binary optimization strategies and how they impact the accuracy of function signature recovery. Specifically, we study the source code of compilers (gcc-8 and clang-7), paying special attention to the mechanism in which arguments are passed from callers to callees under different optimization flags (-O0, -O1, -O2, -O3). We also consult the Intel instruction manual [23] on how each instruction could affect function signatures. Finally, we compile the following eight scenarios in which compiler optimization could impact function signature recovery by the two most recent work, namely TypeArmor and τCFI.

3.3.1 Misidentifying Variadic Functions

As outlined in Sect. 3.2.3, TypeArmor uses $r\dot{w}2s_6()$ as the sole indicator of a variadic function. Interestingly, such a policy tends to introduce more errors in unoptimized binaries in which all arguments are moved onto the stack and any normal function with more than five arguments will be misidentified as variadic. We denote this complication as *Nor2Var*. On the other hand, optimized binaries tend to move arguments to callee-saved registers, which reduces the chances of such errors. That said, normal functions in optimized binaries may still use the stack for parameter passing if the compiler determines that the argument will be reused after the call.

Listing 3.2a shows a function compiled with clang -O2. Since $s_6^{EE} = r\dot{w}2s_6()$, $b2b_5$ is true and $b2b_4$ is false. TypeArmor determines that coff_write_symbol is a variadic function with 4 arguments. However, $|P_{EE-0x471a60}^{GT}| = 7$ as shown at Line 1.

Another complication arises when a variadic function does not use some of the variadic arguments. An optimized binary will not explicitly read these arguments, which will cause the variadic function to be misidentified as normal (denoted as *Var2Nor*). Note that this does not affect binaries compiled by clang since clang always explicitly reads all variadic arguments.

```
 1  static bfd_boolean coff_write_symbol (*,*,*,*,*,*,*)
 2  0000000000471a60 <coff_write_symbol>:
 3  ......
 4  471a6e:      mov      %r9,0x40(%rsp)
 5  471a73:      mov      %r8,0x10(%rsp)
 6  471a78:      mov      %rcx,%r15
 7  471a7b:      mov      %rdx,%r14
 8  471a7e:      mov      %rsi,%rbp
 9  471a81:      mov      %rdi,%r12
10  ......
11  471c4f:      mov      0x40(%rsp),%rbx
```

a Normal function misidentified as variadic

```
 1  void bfd_set_error (bfd_error_type error_tag,...) {
 2      bfd_error = error_tag;
 3      if (error_tag == bfd_error_on_input) {
 4          va_list ap;
 5          va_start (ap, error_tag);
 6          input_bfd = va_arg (ap, bfd *);
 7          input_error = (bfd_error_type)va_arg(ap,int);
 8          ......
 9      }
10  }
11  00000000000328c0 <bfd_set_error>:
12  ......
13  328c4:       mov      %edi,0x300186(%rip)
14  ......
15  328da:       cmp      $0x14,%edi
16  328dd:       mov      %rsi,0x28(%rsp)
17  328e2:       mov      %rdx,0x30(%rsp)
18  328e7:       je       32900
```

b Variadic function misidentified as normal

```
 1  char *concat_copy(char *dst, const char *first, ...)
 2  00000000000dea00 <concat_copy>:
 3  ......
 4  dea25:       test %rsi,%rsi
 5  dea28:       mov %rdx,0x30(%rsp)
 6  dea2d:       mov %rcx,0x38(%rsp)
 7  dea32:       mov %r8,0x40(%rsp)
 8  dea37:       mov %rax,0x8(%rsp)
 9  dea3c:       lea 0x20(%rsp),%rax
10  dea41:       mov %r9,0x48(%rsp)
```

c Number of variadic arguments overestimated

Listing 3.2 Examples of variadic function misidentification

```
 1  GLOBAL(void) jpeg_free_large (j_common_ptr cinfo, void FAR * object, size_t
        sizeofobject) {
 2      free(object);
 3  }
 4  000000000041b6b0 <jpeg_free_large>:
 5  41b6b0: mov    %rsi,%rdi
 6  41b6b3: jmpq   400950 <free@plt>
 7
 8  caller site:
 9  41b5a0: mov    0x70(%r14,%r15,8),%rsi
10  ......
11  41b5d3: mov    %r12,%rdi
12  41b5d6: mov    %rbp,%rdx
13  41b5d9: callq  41b6b0 <jpeg_free_large>
```

Listing 3.3 Not reading argument registers

Listing 3.2b shows a variadic function bfd_set_error compiled by gcc -O2. As shown at Line 6–7, only the first two variadic arguments are used by this function, and therefore gcc only moves %rsi and %rdx onto the stack (Line 16–17). Current approaches would find that $P_{EE-0x328c0}^{OB} = <r\dot{w}_1(32), r\dot{w}2s_2(64), r\dot{w}2s_3(64), c_4, c_5, c_6>$ and determine that $|P_{EE-0x328c0}^{OB}| = 3$ since %r9 is not moved onto the stack. However, $|P_{EE-0x328c0}^{GT}| = 1$ as shown at Line 1.

Moreover, instructions that move the variadic arguments onto the stack in an optimized binary may not be back to back, which results in $b2b$ being unreliable in determining the number of arguments—an overestimation (denoted as **VarOver**). Listing 3.2c shows the variadic function concat_copy compiled by gcc -O2. TypeArmor and τCFI find $b2b_5$ to be false and determine that it is a variadic function with 5 default arguments, but the ground truth is it has only 2 default arguments as shown at Line 1.

3.3.2 Missing Argument-Reading Instructions

When optimization is enabled, there may not be explicit reading of an argument if the function does not use it, leading the corresponding state of the argument to be c. We denote this complication as **Unread**. As shown in Listing 3.3, since the first and third arguments of jpeg_free_large (compiled by clang -O2) are not used, TypeArmor and τCFI determine that $P_{EE-0x41b6b0}^{OB} = <\dot{w}_1(64), r\dot{w}_2(64), c_3, c_4, c_5, c_6>$. Note that compilers always set the argument registers at callers even if they are not used by the callee; see Line 11–12.

```
1  long test(long a, long b)
2  00000000004006a0 <test>:
3  ......
4  4006ae:  callq   400490 <lldiv@plt>
5  4006b3:  mov     %rbx,%rdi
6  4006b6:  mov     %rdx,%rsi
7  4006b9:  callq   *0x200db1(%rip)     # 601470 <fptr3>
8  4006bf:  mov     %rax,%rbx
9  4006c2:  callq   *0x2009a0(%rip)     # 601068 <fptr4>
```

Listing 3.4 Misidentifying %rdx as an argument

3.3.3 Misidentifying %rdx as an Argument

Some registers have special usage in addition to passing arguments. For example, the third argument register %rdx can also be used to store return values when the size of the return value is larger than 64 bits. When there is a read operation on it, current approaches do not distinguish reading an argument from reading the higher 64 bits of a return value. It could then result in an overestimation on the number of arguments. This complication is denoted as *rdx*.

As shown in Listing 3.4, TypeArmor and τCFI determine that $P^{OB}_{EE\ 0x4006a0} = \langle r\dot{w}_1(64), \dot{w}_2(32), r\dot{w}_3(64), c_4, c_5, c_6 \rangle$, and that it is a normal function with 3 arguments. However, $|P^{GT}_{EE-0x4006a0}| = 2$ and the reading of %rdx is for the higher 64 bits of the return value of function lldiv.

3.3.4 Argument (width) Promotion

Some instructions may only work on 64-bit registers or memory, and optimization may prefer using 64-bit registers since using 32-bit registers would result in longer instructions. For example, the compiler uses push to pass arguments to callees (via the stack) when the flag "-mpush-arg" is enabled (e.g., when it is the 7th argument). However, push only allows 64-bit registers as operands, which leads to argument (width) promotion (denoted as *Push*). Line 1–4 of Listing 3.5 shows that the fourth argument out_row_avail, whose type is unsigned int, is passed as the 7th argument at Line 3, and is pushed onto the stack at Line 10 (resulting in $r\dot{w}_4(64)$ instead of $r\dot{w}_4(32)$).

Another complication is due to the default width of operands of certain instructions, e.g., lea [23]. Compilers prefer reading a 64-bit register even if the width of the argument is 32 bits, since reading a 32-bit register requires a prefix 67H (denoted as *lea*). Listing 3.6 shows an example of it, the state of the second argument is $r\dot{w}_2(64)$; however, the ground truth is a 32-bit parameter (unsigned int).

```
 1   typedef unsigned int JDIMENSION;
 2   void process_data_crank_post(j_decompress_ptr cinfo, JSAMPARRAY output_buf,
         JDIMENSION *out_row_ctr, JDIMENSION out_rows_avail) {
 3      (*cinfo->post->post_process_data) (cinfo, NULL, NULL, 0, output_buf,
         out_row_ctr, out_rows_avail);
 4   }
 5   0000000000165c0 <process_data_crank_post>:
 6   165c0:   sub     $0x10,%rsp
 7   165c4:   mov     0x228(%rdi),%rax
 8   165cb:   mov     %rsi,%r8
 9   165ce:   mov     %rdx,%r9
10   165d1:   push    %rcx
11   165d2:   xor     %edx,%edx
12   165d4:   xor     %ecx,%ecx
13   165d6:   xor     %esi,%esi
14   165d8:   callq   *0x8(%rax)
```

Listing 3.5 Promoted argument pushed onto the stack

```
 1   bfd_check_overflow (enum complain_overflow how,
 2   unsigned int bitsize, unsigned int rightshift, unsigned int addrsize, bfd_vma
         relocation)
 3
 4   000000000048ca60 <bfd_check_overflow>:
 5   48ca60:   mov     %ecx,%eax
 6   48ca62:   mov     %edx,%r9d
 7   48ca65:   lea     -0x1(%rsi),%ecx
 8   48ca68:   mov     $0xfffffffffffffffe,%rdx
```

Listing 3.6 Promoted operand of instruction lea

3.3.5   Missing Argument-Writing Instructions

Similar to missing argument reading instructions at callees as discussed above, compiler optimization may decide not to set or reset the value of a register explicitly at callers.

- Higher 64 bits of the return value used as the third argument (denoted as **Ret**). %rdx is used to store the higher 64 bits of the return value. If the compiler finds that a function uses this value as the third argument, it will not explicitly reset %rdx again.
- Uninitialized variable as an argument (denoted as **Uninit**). clang generates undef values for uninitialized variables and do not explicitly set these arguments [24, 25]. On the other hand, gcc initializes them to zero.[3]

---

[3]https://github.com/gcc-mirror/gcc/blob/master/gcc/init-regs.c.

- Indirect calls in wrapper functions (denoted as **Wrapper**). Indirect callers may not reset argument registers when their values are already in the corresponding registers especially for inlined functions.
- Argument values not modified between two calls (denoted as **Unmodified**). gcc-7 and above eliminates writing across functions when the argument register is set to the same value for two consecutive callers.

All the above except **Wrapper** leads to $\hat{w}$ and results in underestimation on the number of arguments. Listing 3.7a presents the higher 64-bit return value and an uninitialized variable are used as arguments. The state vectors for the two indirect calls are $P^{OB}_{ER-0x4006b9} = <w_1(64), w_2(64), \hat{w}_3, \hat{w}_4, \hat{w}_5, \hat{w}_6>$ and $P^{OB}_{ER-0x4006c2} = <\hat{w}_1, \hat{w}_2, \hat{w}_3, \hat{w}_4, \hat{w}_5, \hat{w}_6>$, respectively, which lead to a finding of $|P^{OB}_{ER-0x4006b9}| = 2$ and $|P^{OB}_{ER-0x4006c2}| = 0$. However, by observing the source code at Line 6–7, we realize that $|P^{GT}_{ER-0x4006b9}| = 3$ and $|P^{GT}_{ER-0x4006c2}| = 2$. Listing 3.7b shows indirect calls in a wrapper function. Since there is no direct caller for function bfd_elf64_swap_dyn_in, TypeArmor and τCFI determine that $P^{OB}_{ER-0x416845} = <w_1(64), w_2(64), w_3(64), w_4(64), w_5(64), w_6(64)>$, which results in an overestimation on the number of arguments while $|P^{GT}_{ER-0x416845}| = 1$. Listing 3.7c shows that $P^{GT}_{ER-0x1aae2c} = <w_1(64), w_2(64), w_3(64), \hat{w}_4, \hat{w}_5, \hat{w}_6>$. However, $P^{OB}_{ER-0x1aae2c} = <w_1(64), \hat{w}_2, w_3(64), \hat{w}_4, \hat{w}_5, \hat{w}_6>$ and $|P^{OB}_{ER-0x1aae2c}| = 1$ since the value of %rsi is not changed by the function at 0x1a95f0, and the compiler does not reset it explicitly.

3.3.6 Registers Storing Temporary Values

Since all argument registers are general-purpose registers, they could also be used as scratch registers to store temporary values, which could result in an overestimation on the number of arguments (denoted as **Temp**). Listing 3.8a shows an example (compiled with clang -O0) with $P^{OB}_{ER-0x416015} = <w_1(64), w_2(64), w_3(64), w_4(64), \hat{w}_5, \hat{w}_6>$ and $|P^{OB}_{ER-0x416015}| = 4$. However, according to the ground truth at Line 7, we can observe that $|P^{GT}_{ER-0x416015}| = 3$ and the write operation on %rcx is to store a temporary value. Note that compiler optimization can remove many redundant instructions that are used to store temporary values; and so it has a positive impact on this case; see the optimized binary in Listing 3.8b where $P^{OB}_{ER-0x438891} = <w_1(64), w_2(64), w_3(64), \hat{w}_4, \hat{w}_5, \hat{w}_6>$ and $|P^{OB}_{ER-0x438891}| = |P^{GT}_{ER-0x438891}| = 3$.

3.3.7 Argument (width) Demotion

To the opposite of argument promotion at callees, compilers may use a smaller-sized register (32-bit), since a 64-bit register may need a REX prefix [23] which increases the code size and affects the I-cache footprint. This applies to cases where

```
1  long test2(long a, long b){
2      //mesg and err are not initialized
3      char *mesg,*err;
4      lldiv_t res;
5      res = lldiv (31558149LL,3600LL);
6      long r1 = (*fptr3)(a, res.quot, res.rem);
7      (*fptr4)(mesg,err);
8      printf("%s\n", buffer);
9      return r1;
10 }
11 00000000004006a0 <test2>:
12 ......
13 4006ae: callq   400490 <lldiv@plt>
14 4006b3: mov     %rbx,%rdi
15 4006b6: mov     %rax,%rsi
16 4006b9: callq   *0x200db1(%rip) # 601470 <fptr3>
17 4006bf: mov     %rax,%rbx
18 4006c2: callq   *0x2009a0(%rip) # 601068 <fptr4>
19
```

a Missing argument-writing instructions

```
1  0000000000416830 <bfd_elf64_swap_dyn_in>:
2  416830: push    %r15
3  ......
4  416835: mov     %rdx,%r14
5  416838: mov     %rsi,%r15
6  41683b: mov     %rdi,%rbx
7  41683e: mov     0x8(%rdi),%rax
8  416842: mov     %rsi,%rdi
9  416845: callq   *0x68(%rax)
10
```

b An indirect call in a wrapper function

```
1  1aae0a: mov     0xb38(%r13,%r14,1),%rdi
2  1aae12: mov     %rbp,%rsi
3  1aae15: callq   1a95f0
4  1aae1a: mov     (%rsp),%rax
5  1aae1e: lea     (%rax,%r14,1),%rdx
6  1aae22: mov     (%rbx),%rax
7  1aae25: mov     0xb8(%rax),%rdi
8  #call funcs->create( cffsize->face->memory, &priv, &internal->subfonts[i −
        1] )
9  1aae2c: callq   *(%r12)
10
```

c Arguments not modified between two calls

Listing 3.7 Missing argument-writing instructions

```
1  ......
2  460ffc:  mov    -0x18(%rbp),%rdi
3  461000:  mov    -0xe8(%rbp),%rsi
4  461007:  mov    -0xf0(%rbp),%rcx
5  46100e:  add    $0x10,%rcx
6  461012:  mov    %rcx,%rdx
7  #(*bed->elf_backend_reloc_type_class)(info, o, s->rela);
8  461015:  callq  *%rax
```

a Assembly compiled with clang -O0.

```
1  ......
2  438881:  mov    0x30(%rsp),%rdi
3  438886:  mov    %rbx,%rsi
4  438889:  mov    %rbp,%rdx
5  43888c:  mov    0x10(%rsp),%rax
6  #(*bed->elf_backend_reloc_type_class)(info, o, s->rela);
7  438891:  callq  *0x208(%rax)
```

b Assembly compiled with clang -O2.

Listing 3.8 Registers to store temporary values

- Arguments are constants whose sizes are up to 32 bits (denoted as ***Imm***);
- Arguments are pointers pointing to .rodata, .bss, and .text sections (denoted as ***Pointer***); and
- Arguments are NULL pointers (denoted as ***Null***).

Listing 3.9a shows an example for these cases compiled by clang -O2. The ground truth at Line 4 shows $P_{ER-0x546593}^{GT} = <w_1(64), w_2(64), w_3(64), \hat{w}_4, \hat{w}_5, \hat{w}_6>$, while TypeArmor and τCFI determine that $P_{ER-0x546593}^{OB} = <w_1(64), w_2(32), w_3(32), \hat{w}_4, \hat{w}_5, \hat{w}_6>$ since the second argument (0x8a01b0) is a pointer pointing to the .rodata section, and the third argument (0x2000) is a 32-bit constant. The example with a NULL pointer being an argument is shown in Listing 3.9b. According to the ground truth at Line 5, the second argument should be a pointer; but a NULL pointer is passed at the caller, and the compiler uses xor to prepare for it.

3.3.8 Argument (width) Promotion at Both Callees Andcallers (Prom)

There are other argument (width) promotions at both callees and callers that would not result in inaccuracies in matching function callees with callers since the argument

```
1  546586:mov  $0x8a01b0,%esi
2  54658b:mov  $0x2000,%edx
3  546590:mov  %r14,%rdi
4  #(*git_hash_update_fn)(*, *, size_t len);
5  546593:  callq  *0x28(%rax)
```

a A constant and a pointer as arguments

```
1  57f50e:test  %rbp,%rbp
2  57f511:je  57f531
3  57f513:mov  0x333a46(%rip),%rdi
4  57f51a:xor  %esi,%esi
5  #(*advertise)(*r, *);
6  57f51c:  callq  *0x8(%rbp)
```

b A NULL pointer as an argument

Listing 3.9 Argument width demotion

promotion happens in a matching manner. This refers to promotions of types smaller than the native type of the target platform's Arithmetic Logic Unit (ALU) to make arithmetic and logical operations possible or more efficient. C and C++ perform such promotions for objects of boolean, character, wide character, enumeration, and short integer types. As shown in Listing 3.10, the type of the third argument is unsigned char (8-bits) as shown at Line 1, but the analysis engine would determine its state being $riw_3(32)$ due to the promotion performed by the compiler.

Table 3.1 presents a summary on the complications at both callees and callers with the last column indicating the consequences.

3.4   Experimental Results of the Eight Complications on Real-World Programs

Section 3.3 details our theoretical analysis by analyzing compiler optimization strategies. In this section, we test how the eight complications identified in Sect. 3.3 present themselves in real-world programs. Specifically, we use a test suite of programs comprising of 552 C and 792 C++ applications compiled with gcc-8 and clang-7 with optimization levels from -O0 to -O3 for x86-64, and compare analysis results of TypeArmor and τCFI with ground truths extracted. Since the source code of τCFI is not released, we implement it ourselves according to the description of the paper [18].

Table 3.1 Summary of complications introduced by compiler optimization

Site	Category	Complication	Impact
Callee	Misidentifying variadic functions	Normal to variadic (*Nor2Var*)	$\lvert P_{EE}^{OB}\rvert < \lvert P_{EE}^{GT}\rvert$
		Variadic to Normal (*Var2Nor*)	$\lvert P_{EE}^{OB}\rvert > \lvert P_{EE}^{GT}\rvert$
		Back-to-back condition unreliable (*VarOver*)	$\lvert P_{EE}^{OB}\rvert > \lvert P_{EE}^{GT}\rvert$
	Missing argument reading instructions	Arguments are not used by a function (*Unread*)	$\lvert P_{EE}^{OB}\rvert < \lvert P_{EE}^{GT}\rvert$ $\lvert s_i^{EE}\rvert < \lvert s_i^{EE,GT}\rvert$
	Misidentifying %rdx as an argument	Reading the higher 64 bits of a return value (*rdx*)	$\lvert P_{EE}^{OB}\rvert > \lvert P_{EE}^{GT}\rvert$
	Argument (width) promotion	Arguments are pushed onto the stack (*Push*)	$\lvert s_i^{EE}\rvert > \lvert s_i^{EE,GT}\rvert$
		Default width of the operand of certain instructions is 64-bit (*lea*)	$\lvert s_i^{EE}\rvert > \lvert s_i^{EE,GT}\rvert$
Caller	Missing argument writing instructions	Higher 64 bits of a return value as the third argument (*Ret*)	$\lvert P_{ER}^{OB}\rvert < \lvert P_{ER}^{GT}\rvert$
		Uninitialized variables as arguments (*Uninit*)	$\lvert P_{ER}^{OB}\rvert < \lvert P_{ER}^{GT}\rvert$
		Indirect calls in wrapper functions (*Wrapper*)	$\lvert P_{ER}^{OB}\rvert > \lvert P_{ER}^{GT}\rvert$
		Argument values not modified between two calls (*Unmodified*)	$\lvert P_{ER}^{OB}\rvert < \lvert P_{ER}^{GT}\rvert$
	Registers storing temporary values	Argument registers are used to store temporary values (*Temp*)	$\lvert P_{ER}^{OB}\rvert > \lvert P_{ER}^{GT}\rvert$
	Argument (width) demotion	Argumets are constant whose sizes are up to 32-bit (*Imm*)	$\lvert s_i^{ER}\rvert < \lvert s_i^{ER,GT}\rvert$
		Argument are pointers pointing to data and text sections (*Pointer*)	$\lvert s_i^{ER}\rvert < \lvert s_i^{ER,GT}\rvert$
		Arguments are NULL pointers (*Null*)	$\lvert s_i^{ER}\rvert < \lvert s_i^{ER,GT}\rvert$
Both	Small integral type promotion	Small integral types are promoted to native types (*Prom*)	$\lvert s_i^{EE}\rvert > \lvert s_i^{EE,GT}\rvert$ $\lvert s_i^{ER}\rvert > \lvert s_i^{ER,GT}\rvert$

```
1        static bfd_boolean add_line_info (struct line_info_table *table,
         bfd_vma address, unsigned char op_index, char *filename,
         unsigned int line, unsigned int column, unsigned int discriminator,
         int end_sequence)
2
3        000000000044c2d0 <add_line_info>:
4        44c2d0: push    %rbp
5        ......
6        44c2e7: mov     %edx,%r12d
7        44c2ea: mov     %rsi,%r13
8        44c2ed: mov     %rdi,%rax
9        44c2f0: mov     (%rdi),%rdi
10       44c2f3: mov     %rax,0x8(%rsp)
11       44c2f8: mov     0x30(%rax),%rax
12       44c2fc: mov     %rax,0x10(%rsp)
13       44c301: mov     $0x28,%esi
14       44c306: callq   408a80 <bfd_alloc>
15
```

Listing 3.10 Promotion of small integral types

In addition to TypeArmor and τCFI which recover function signatures for the specific purpose of Control-Flow Integrity, we also include a well-known binary analysis framework, Ghidra [21] v9.1.1, into our experiments since it also performs function signature recovery for reverse engineering purposes. Besides its general-purpose nature which leads to less emphasis on precision of the function signature recovery, our preliminary analysis on its source code reveals the following distinctions when Ghidra is compared to TypeArmor and τCFI in their mechanisms of function signature recovery:

- Only functions with symbol information are correctly identified as variadic, while those without symbol information are simply assumed to be non-variadic;
- Only instructions immediately prior to (without control-flow transfers) a call instruction are considered potentially preparing for function arguments;
- Forward and backward analysis are constrained within the scope of a single function; and
- Width for each argument at callers is always 64 bits.

With this preliminary understanding, we expect Ghidra to perform less accurately compared to TypeArmor and τCFI in recovering function signatures.

Our test suite is composed of Binutils-2.26, LLVM test-suite, and C/C++ applications from Github. This composition ensures that (1) it contains a wide variety of realistic C and C++ binaries with sizes ranging from 0.07MB to slightly more than 100MB (see Table 3.2 for details of sizes of the binary executables); (2) it contains binaries used in the evaluation of previous work, making it possible to compare our results with the literature; (3) it includes real-world applications downloaded from Github which contain complex corner cases which "testbed" applications may not

Table 3.2 Sizes of the binary executables in our test suite

Language	Opt	Size (MB)					
		clang			gcc		
		min	median	max	min	median	max
C	O0	0.07	0.69	44.75	0.08	0.68	44.72
	O1	0.07	0.71	45.61	0.12	0.98	50.52
	O2	0.08	0.84	50.09	0.11	1.02	51.79
	O3	0.08	0.84	48.95	0.13	1.55	54.30
C++	O0	0.11	7.51	65.77	0.12	14.60	73.22
	O1	0.11	7.22	68.82	0.17	10.32	99.959
	O2	0.13	6.31	65.70	0.18	16.96	105.50
	O3	0.13	6.15	66.79	0.19	17.12	109.83

Table 3.3 Github applications in our test suite

App	Language	Description
git	C	Distributed version control system
darknet	C	An open source neural network framework
netdata	C	A real-time performance monitoring
redis	C	An in-memory database
sqlite	C	SQL database engine
vim	C	UNIX text editor
gnupg	C	Complete implementation of the OpenPGP standard
openssl	C	TLS/SSL and crypto library
mupdf	C & C++	A lightweight PDF, XPS, and E-book viewer
vorbis	C	A general purpose audio and music encoding format
aria2c	C++	A lightweight multi-protocol download utility
cppcheck	C++	Static analysis of C/C++ code
hpx	C++	C++ Standard Library for Parallelism and Concurrency
xpdf	C++	A PDF viewer and toolkit

have (see Table 3.3 for details of the Github applications we choose—mainly those with many "stars").

3.4.1 Ground Truth and Statistics on the Ground Truth

Our objective of the experiments is to compare results from TypeArmor, τCFI, and Ghidra with ground truths to see how the complications identified in Sect. 3.3 present

Table 3.4 Number of arguments of functions in our test suite

Language	Opt	Number of Arguments (%)							# variadic
		0	1	2	3	4	5	6	
C	O0	6.92	29.35	29.73	17.46	7.47	4.33	1.77	8.45
	O1	6.01	28.64	29.87	17.32	7.73	4.71	1.95	
	O2	6.78	28.05	27.85	18.11	8.22	4.92	1.88	
	O3	5.99	26.52	29.04	18.20	8.55	5.17	2.11	
C++	O0	4.31	47.84	26.78	12.97	3.64	2.47	0.64	2.43
	O1	4.44	46.06	27.76	13.34	3.80	2.54	0.67	
	O2	3.09	45.27	20.77	12.58	7.09	5.48	1.87	
	O3	3.13	45.84	20.88	12.45	6.95	5.09	1.86	

themselves in real-world applications. Here we first briefly explain how we obtain the ground truth in an automatic manner.

We base our ground truth on information collected by an LLVM [26] pass and on DWARF v4 debugging information [27] which is the default setting for gcc and clang. We use LLVM to collect source-level information, including the number and types of arguments for each function and indirect callers when the arguments are integers (using LLVM API isIntegerTy(N)) and pointers (using LLVM APIs isPointerTy() and isFunctionTy()[4]). We also record the source line numbers of functions and indirect callers. We then compile the test applications with DWARF information and link the source-level line numbers with binary-level addresses using the DWARF line number table.

We implement the above with more than 500 lines of C++ code and more than 2,000 lines of python code. The result is a ground truth file for each binary in the test suite. With the ground truth collected, we perform statistical tests on our test suite to ensure that applications included could potentially present all variety of function signatures. Specifically, we count the number of arguments (ground truth) of all functions and make sure that there are sufficient numbers of functions with the number of arguments from 0 to 6; see Table 3.4 for details. We observe that there are more functions with between 1 and 3 arguments, and that C programs are more likely to have variadic functions. We also check the (ground truth) argument types for each function (see Table 3.5, which shows the percentage of functions having a specific type as its arguments). It appears that pointers are heavily used as function arguments, especially for C++ applications. This may imply that C++ applications are less likely to present complications on argument width demotion or promotion.

---

[4]We also check whether a struct argument has the attribute ByVal since clang will copy it onto the stack while considering it as a pointer.

Table 3.5 Argument types of functions in our test suite

Type	Opt	Arg for C (%)						Arg for C++ (%)					
		1st	2nd	3rd	4th	5th	6th	1st	2nd	3rd	4th	5th	6th
8-bits	O0	0.19	0.37	0.12	0.26	0.34	0.35	0.02	0.32	1.52	0.27	0.83	0.94
	O1	0.14	0.31	0.23	0.29	0.54	0.56	0.08	0.56	1.60	0.38	0.43	0.55
	O2	0.11	0.24	0.30	0.23	0.75	0.78	0.02	0.31	1.48	0.26	0.79	0.92
	O3	0.10	0.25	0.26	0.29	0.60	0.61	0.02	0.32	1.52	0.27	0.83	0.94
16-bits	O0	0.09	0.19	0.13	0.14	0.17	0.29	-	0.04	0.22	0.24	0.44	0.50
	O1	0.11	0.24	0.17	0.12	0.13	0.27	0.04	0.08	0.11	0.22	0.39	0.31
	O2	0.10	0.21	0.10	0.16	0.14	0.33	-	0.04	0.22	0.23	0.42	0.49
	O3	0.11	0.24	0.15	0.13	0.0.14	0.29	-	0.04	0.22	0.24	0.44	0.50
32-bits	O0	9.38	19.58	25.33	29.65	33.50	28.79	0.82	9.66	19.57	26.02	29.27	17.36
	O1	8.55	19.75	25.29	30.31	37.60	32.97	0.55	4.60	15.19	15.57	21.03	16.80
	O2	8.31	18.41	24.21	28.38	32.71	25.61	0.81	9.44	19.10	24.89	27.83	16.98
	O3	7.48	19.23	24.91	30.48	38.84	33.33	0.82	9.66	19.57	26.02	29.27	17.36
64-bits	O0	2.31	7.25	11.36	10.37	10.85	10.00	0.14	5.13	10.82	17.14	13.03	10.47
	O1	1.97	6.17	10.93	9.18	9.01	7.74	0.83	7.54	14.87	11.21	7.37	6.29
	O2	2.24	6.99	12.19	11.08	10.77	9.83	0.14	4.96	10.92	17.21	13.04	10.31
	O3	1.94	5.95	10.65	9.31	8.69	7.57	0.14	5.13	10.82	17.14	13.03	10.47
ptr	O0	88.02	72.63	62.61	59.43	54.99	60.44	98.03	84.45	67.56	54.70	56.15	70.21
	O1	89.22	73.40	62.94	59.84	52.48	58.22	98.28	86.99	67.61	70.73	70.20	75.77
	O2	89.24	73.96	62.77	59.79	55.45	63.39	98.06	84.87	67.99	55.85	57.66	70.80
	O3	90.36	74.14	63.58	59.48	51.48	57.95	98.03	84.45	67.56	54.70	56.15	70.21

3.4.2 Metric Used and Overall Results

Since applications may have different numbers of functions and functions may have different numbers of indirect callers, we do not directly calculate the geometric mean as in TypeArmor [19] and τCFI [18]. Instead, we calculate the geometric mean of the *likelihood* that the callees and indirect callers present a complication in their function signature recognition. Specifically, we calculate the likelihood that the complications discussed in Sect. 3.3 cause under- and overestimation on the recovered function signatures. For example, application addr2line compiled with clang -O0 has 2,019 normal functions among which 101 are misidentified as variadic and the identified number of arguments is underestimated. We first calculate the likelihood that a function is misidentified in this application (101/2019), and

Table 3.6 Number of functions and indirect calls

		Opt	#func	#normal func	#variadic	#icalls
clang	C	O0	543	486	21	64
		O1	540	495	19	60
		O2	394	346	19	85
		O3	380	352	19	93
	C++	O0	3,379	3,229	15	121
		O1	3,290	3,085	13	74
		O2	702	652	13	439
		O3	710	640	13	452
gcc	C	O0	546	483	24	70
		O1	446	420	19	69
		O2	418	386	21	67
		O3	406	370	22	83
	C++	O0	4,505	4,113	22	152
		O1	686	608	13	336
		O2	698	613	13	312
		O3	656	606	13	299

then use this number to compute the geometric mean for all applications in our test suite; see Fig. 3.1[5] and Fig. 3.2.[6]

We discuss the detailed findings in the following sections. Note that complication case *Unmodified* only appears in one application (mupdf[7] compiled with gcc) and that *Uninit* and *Ret* do not appear at all in our test suite. We stress that this does not indicate insufficiency in our experiment, but rather the complications identified in our theoretical analysis (Sect. 3.3) do not necessarily present themselves in real-world programs.

Unread: This is by far the biggest contributor to misidentification of function signatures at callees, where the fact that many functions do not read (some of) their arguments leads to underestimation of the number of arguments. It also potentially leads to underestimation of the width of an argument register whose evident reading instruction is missing while existence is implied (due to subsequent argument registers whose reading instructions being present). This complication presents more heavily in C++ programs due to the simplicity of many (callee) functions whose

---

[5]Likelihood is calculated against the number of normal functions for *Nor2Var*, against the number of variadic functions for *Var2Nor* and *VarOver*, against the total number of functions for *rdx*, *Unread*, *Push*, *lea* and *Prom*. See Table 3.6 for the number of various types of functions in our test suite.

[6]Likelihood is calculated against the total number of indirect calls. See Table 3.6 for the number of indirect calls in our test suite.

[7]https://mupdf.com/.

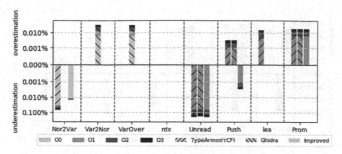

(a) C applications compiled by Clang

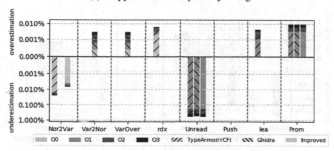

(b) C++ applications compiled by Clang

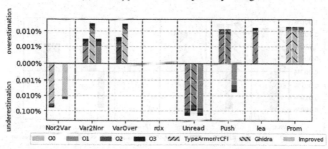

(c) C applications compiled by GCC

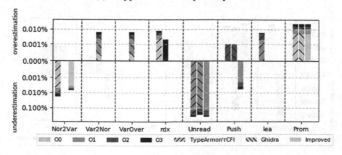

(d) C++ applications compiled by GCC

Fig. 3.1 Likelihood of complications at callees

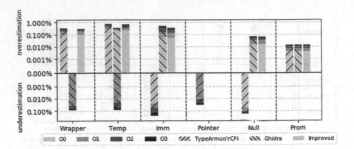

(a) C applications compiled by Clang

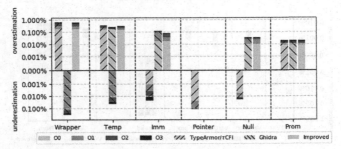

(b) C++ applications compiled by Clang

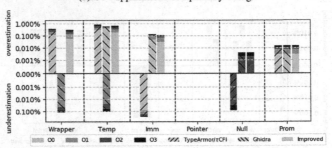

(c) C applications compiled by GCC

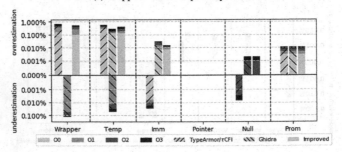

(d) C++ applications compiled by GCC

Fig. 3.2 Likelihood of complications at callers

implementation does not require accessing the *this* argument. Another finding is that C++ applications compiled by gcc tend to have dead code eliminated, which makes them seemingly less vulnerable to this complication. Note that unoptimized binaries do not have this issue at all because compilers always insert argument reading instructions even if the callee function does not need them.

Nor2Var: This also presents heavily in our test suite, leading to underestimation on the number of arguments, especially in C programs, except that compiler optimization actually helps mitigating it. As explained in Sect. 3.3.1, unoptimized binaries always move all arguments onto the stack, making it more likely to present more than 5 integer arguments at the callee which always leads to misidentification of variadic functions. Optimization helps "skipping" some of the arguments and reducing the likelihood of misidentification. Ghidra is immune to this complication since it simply considers all functions non-variadic.

lea, *Push*, and *Prom*: These three complications result in overestimation on the argument width, and together present a large thread to function signature identification of optimized binaries. Checking into the details, we find that C programs make heavier use of lea to perform simple computations and more often push arguments onto the stack (especially with gcc). Looking into the case of *Prom*, we find that clang -O0 does not promote the argument width (it uses register al or ax to store the argument) while gcc does (it uses eax) even when optimization is turned off.

rdx: This presents more on C++ programs and leads to overestimation on the number of arguments. Upon checking the details, we realize that the exception handling in C++ will call function rethrow_exception, which invokes function _Unwind_RaiseException that returns the unwind reason code in %rdx and the exception object in %rax.

Var2Nor: As expected, Ghidra is vulnerable to this, although not that much due to compiler optimization but the simple treatment it employs (all functions are non-variadic). This complication presents to TypeArmor and τCFI, and is usually due to empty implementation of functions with more than five compulsory arguments. We find that C programs compiled by gcc suffer overestimation on the number of arguments on top of function type misidentification.

VarOver: This only presents itself on binaries compiled with gcc -O2 and -O3, where the instructions that move the variadic arguments onto the stack are not back to back. On the other hand, all variadic functions are identified as non-variadic in Ghidra, so the number of arguments is overestimated.

Temp and *Wrapper*: These are clear examples in which compiler optimization helps TypeArmor and τCFI determining the number of function arguments. In the case of *Temp*, optimization eliminates redundant instructions as function arguments. *Wrapper* causes fewer complications in optimized binaries due to heavier applications of function inlining. Note that C++ applications are more vulnerable to *Wrapper* due to the large number of virtual functions being called indirectly. Ghidra generally

Table 3.7 Likelihood that indirect calls in C programs use immediate values as arguments

Compiler	Opt	Non-inline	Inline	Loop-unroll	Func-copy
clang	O0	24	0	0	3
	O1	23	0	0	6
	O2	13	52	0	3
	O3	13	49	45	4
gcc	O0	81 (25)	0	0	3
	O1	75 (23)	59	0	5
	O2	26	28	0	3
	O3	22	33	3	3

Numbers in brackets correspond to functions that pass the value 0 to an argument register.
"Func copy" refers to multiple copies of the same function called from different modules.
Results shown are likelihood results multiplied by 1,000, rounded to the nearest integer

performs worse here (considering the combined errors in both over- and underestimation) mainly due to its limited scope of backward analysis for indirect calls in wrapper functions. That said, Ghidra has superior mechanisms in dead code elimination and only the basic block which contains an indirect call is analyzed, which results in some argument registers that are used for temporary storage being correctly identified; see the complication of *Temp* (overestimation).

Imm **and** *Null*: C applications compiled with clang and gcc are both likely to pass immediate values to argument registers, which results in underestimation of the argument width by TypeArmor and τCFI. Interestingly, the likelihood increases upon increase of optimization levels. Digging into the details, we realize that this is actually just an artifact because higher optimization level results in heavier application of function inlining (-O1 and -O2 for clang, -O1, -O2, and -O3 for gcc) and loop unrolling (-O3 for both compilers), which leads to a larger number of callers of the same function; see Table 3.7 which shows the likelihood that indirect calls use immediate values as arguments for different reasons. Another interesting observation is that gcc -O0 and -O1 are more likely to move zero (*Null*) to an argument register than using xor.

Ghidra, on the other hand, is not vulnerable to this underestimation but rather suffers on overestimation because it always uses the entire 64-bit memory range as the argument width.

Pointer: This only affects applications compiled with clang especially on C++ programs as they are more likely to pass pointers to indirect callees. C programs compiled with clang -O0 do not have this problem because it uses a 64-bit register to store the pointer by adding a prefix to denote the use of a 64-bit displacement or immediate source operand. C++ programs, on the other hand, set a 32-bit register to the pointer address and then move it to the argument register for some indirect calls. We also find that C++ applications compiled with clang -O1 have a higher likelihood on this complication. This is because for some indirect calls that accept pointers as arguments, `clang` prepares them by moving 64-bit immediate values onto the stack first, and then after another indirect call instruction, the argument register is set by reading the 64-bit value from this stack address. As the number of indirect calls in binaries compiled with -O2 and -O3 is much larger, the likelihood for them becomes smaller.

Ghidra, again, is not vulnerable to this because it always uses the entire 64-bit memory range as the argument width.

Applications compiled by `gcc` do not use pointers that point to .text, .rodata, or .bss as arguments because `gcc-7` and above compile applications into position-independent code.

Prom: This seems to be less sensitive to compiler optimization (compiler will always promote to the native type—32 bits) and only affects a small number of indirect calls.

3.5   Our Compiler-Optimization-Friendly Policies

In an effort to properly handle the complications arisen due to compiler optimizations to more accurately recover function signatures, we propose a set of improved policies. In this section, we first discuss the details of these policies and then present our evaluation results of applying them to analyze our test suite of 1,344 real-world applications. Note that most of the policies proposed here are generally accurate for both optimized and unoptimized binaries, while others are more specifically targeting optimized binaries. Existing work [28, 29] and our experience (e.g., if values of all six argument registers are moved onto the stack, then it must be an unoptimized binary) show that detecting the compiler and the optimization level used in well-behaved binaries can be done accurately, and we take it as a prerequisite of enforcing our policies specifically targeting optimized binaries.

Identifying Variadic Functions (Targeting *Nor2Var* and *VarOver*)
The main problem in existing approaches is the identification of variadic arguments using "back-to-back value assigning instructions" (i.e., *b2b*) [19], which is not a sufficient condition as we analyzed (see Sect. 3.3.1) and showed in experiments (see Sect. 3.4). We discover another more direct and sufficient condition for variadic argument identification when optimization is enabled, in which the stack addresses storing variadic arguments are consecutive, prepared using 64-bit registers, and read using pointers. More specifically,

Definition 12. *Let $@_i$ denote the stack address to which argument register i is moved given $r\dot{w}2s_i()$. Callee function f is a variadic function iff $\forall i \in \{5, 4, 3, 2, 1\}$,*

- $|@_{i+1} - @_i| = 8;$ *and*
- $s_{i+1}^{EE} = r\dot{w}2s_{i+1}(64)$ *and* $s_i^{EE} = r\dot{w}2s_i(64);$ *and*
- $@_{i+1}$ *and* $@_i$ *are read via pointers.*

Table 3.8 Analysis of the non-variadic function in Binutils

Line number	Operation	TypeArmor	Our improved policy
4	Move %r9 to stack 0×40(%rsp)	%r9 is a variadic argument	May be a variadic argument
5	Move %r8 to stack 0×10(%rsp)	%r8 is a variadic argument	Non-consecutive stack addresses; not a variadic argument
11	0×40(%rsp) is read not overwritten		Not a variadic argument
Conclusion		Variadic function with 4 arguments	Normal function with 6 arguments

with $|P_{EE-f}^{OB}|$ being the maximal i violating the above. Otherwise, f is a normal function and $|P_{EE-f}^{OB}|$ is:

$$\begin{cases} 6 & \text{if } r\dot{w}2s_6() \text{ and } @_6 \text{ is not read via a pointer} \\ \max(\underset{i}{\text{argmax}}(r\dot{w}2s_i()), \underset{i}{\text{argmax}}(r\dot{w}_i())) & \text{if } s_6^{EE} \neq r\dot{w}2s_6() \end{cases}$$

We use the example in Listing 3.2a to show how our policy works. During analysis, we find that $P_{EE-0x471a60}^{OB} = <r\dot{w}_1(64), r\dot{w}_2(64), r\dot{w}_3(64), r\dot{w}_4(64), rw2s_5(64), rw2s_6(64)>$ and $@_6$ is not read via a pinter; therefore, we conclude that $|P_{EE-0x471a60}^{OB}| = 6$. Note that although $|P_{EE-0x471a60}^{GT}| = 7$, $|P_{EE-0x471a60}^{OB}| = 6$ is an accurate and best approximation based on the limited information present in the binary. The details about the analysis result by TypeArmor and our new policy can be found in Table 3.8.

The policy described above does not work well when optimization is disabled, in which all arguments are copied onto the stack at consecutive addresses. The policy to deal with unoptimized binaries is described in Definition 13.

Definition 13. *Callee function f is a variadic function iff* $\forall i \in \{5, 4, 3, 2, 1\}$,

- $|@_{i+1} - @_i| = 8;$ and
- $s_{i+1}^{EE} = r\dot{w}2s_{i+1}(64)$ and $s_i^{EE} = r\dot{w}2s_i(64)$.

with $|P_{EE-f}^{OB}|$ being the maximal i violating the above. Otherwise, f is a normal function with 6 arguments.

3.5.1 Argument (Width) Promotion and Demotion (Targeting Push, lea, Imm, Pointer, and Null)

Our improved policy solves the argument promotion and demotion complications by analyzing the context of the instructions. More specifically,

- **Push**: Let $p = 32$ in $r\hat{w}_i(p)$ if the corresponding argument reading instruction is push.
- **lea**: Let p in $r\hat{w}_i(p)$ be the minimum of the width of the source and destination registers (instead of that of the source only as in TypeArmor and τCFI).
- **Imm**: Let $p = 64$ in $r\hat{w}_i(p)$ if register i holds a constant.
- **Pointer**: Let $p = 64$ in $r\hat{w}_i(p)$ if register i holds a pointer value pointing to .rodata, .bss, or .text section.
- **Null**: Let $p = 64$ in $r\hat{w}_i(p)$ if register i is involved in an xor instruction.

Note that this improved policy guarantees that all legal callers be matched with legal callees since there is no underestimation at callers or overestimation at callees, but could lead to some imprecise (but conservative) results. For example, demoting the argument width to 32 bit for a register read using push may result in underestimation; see the case of **Push** in Fig. 3.1. We believe that this is a good tradeoff where an absolutely precise solution does not exist, especially since the intended control flow is never broken with our improved CFI policy.

3.5.2 Register Overloading (Targeting rdx)

Since the overloading of rdx is for storing function return values, we simply consider any first reading of %rdx after a call to a library function (let's denote the callee f) as $w_3()$. It may first sound counter-intuitive, but this must be reading the return value of f since the compiler has to make a conservative assumption that f has reset %rdx. This improved policy solves the complication **rdx** at callees with 100% accuracy.

3.5.3 Registers Storing Temporary Values (Targeting Temp)

Recall that the analysis of callers considers all instructions involving an argument-passing register instead of focusing on only the first instruction (Sect. 3.2). Although that is technically correct, it also introduces complications since registers storing temporary values could be miscounted as passing parameters to a callee (**Temp**). Our improved policy takes into consideration the reading of registers (rather than focusing only on writing in the original policy) as well as the sequence of the instructions. More specifically, we let $s_i^{ER} = \hat{w}_i$ if register i is moved to another argument register after the write operation when the value of register i is not zero (a special case where the compiler will directly move register i to another argument register since the compiler does not prefer passing zeros to a register directly).

For example, as shown in Listing 3.8a, %rcx is moved to %rdx at Line 6 after the write operation at Line 4. With this, we conclude that %rcx is not used to pass arguments and $|P^{OB}_{ER-0x461015}| = 3$.

In order to be conservative, we only apply this policy to basic blocks where indirect calls are located. Note that this policy can also help correctly recover the number of arguments for indirect calls in wrapper functions.

3.5.4   Additional Binary Analysis to Extract Our Policies

We have presented *what* our improved CFI policies are so far in this section. Here we briefly discuss *how* it is done with the additional binary analysis we perform.

Our improved policy for *Nor2Var* requires that we trace the data flow of a stack memory to check whether it is read without being overwritten. This is done by following the CFG of a function and check whether the stack memory is used as the source operand without being used as a destination operand.

Our improved policy for *Imm* requires that we identify whether one register holds a constant. Specifically, during the backward analysis, if we encounter a 32-bit argument register being written to, we will record its source recursively and check whether it is an immediate value. Our experiences show that this recursive tracing typically reports a success within the same basic block and does not result in excessive overhead.

3.5.5   Evaluation of Our Improved Policies

We apply our new policies on the same test suite consisting of 1,344 C and C++ applications and use the same metric as described in Sect. 3.4.2 to evaluate it; see the bars named "Improved" in Fig. 3.1 and Fig. 3.2. The comparison shows that our new policies result in significant improvement over most of the complication cases. In particular, we completely mitigate the complication cases of *VarOver*, *rdx*, *lea*, and *Pointer*, and significantly reduce the chances of running into *Nor2Var*.

For cases of *Imm*, *Null*, and *Push*, our policy guarantees that valid calls are never inadvertently blocked, but it could also potentially make the recovered function signatures more conservative. For example, we promote the argument width at indirect callers for cases *Imm* and *Null*, which may result in overestimation on argument widths as shown in Fig. 3.2 with likelihood less than 10.1% and 1.7%, respectively. Similarly, our policy to deal with *Push* may cause argument width underestimation at the callees, and the likelihood is about 0.2%. This raises an interesting question whether it is possible for CFI policies recovered from binary executables to be more accurate and approach the accuracy of source-based solutions; we discuss this in Sect. 3.6.1.

For *Nor2Var*, the likelihood of misidentifying normal functions to variadic for unoptimized binaries is reduced from 3.3% to 1.2%, with that for optimized binaries dropped to 0.1%.

Since we only apply the policy for *Temp* to basic blocks where indirect calls are located, there can be overestimations if the argument registers storing temporary values are in other predecessors. The same policy also helps identify the number of arguments for indirect calls in wrapper functions as shown in the case of *Wrapper* in Fig. 3.2—the likelihood of overestimation on the number of arguments is reduced from 11.5% to 5.4% for C applications compiled by gcc -O0.

3.5.6 Potential Revisions to Deal with Other Complications

To handle *Var2Nor*, we could revise our policy on identifying variadic functions to find the argument register with the highest index i that is moved onto the stack. However, this will result in (potentially unnecessary) checking of registers at a smaller index, and lead to substantially higher overhead in the processing. Since we only observe one variadic function (bfd_set_error in Binutils) being misidentified as a normal function and causing overestimation on the number of arguments in our large test suite, we do not suggest enforcing this policy.

Similarly for *Unmodified*, we could perform backward analysis from the indirect caller until another indirect call is encountered. We do not enforce this policy because there is only one application in our test suite that has this problem (with only two indirect calls), and this policy could result in a large number of overestimation on the number of arguments at indirect callers.

3.6 Discussions and Security Implications

In this section, we first discuss an interesting question whether policies recovered from binary executables could approach the accuracy of source-based solutions, and then further evaluate the security implications of having inaccurate CFI policies.

3.6.1 Comparison with Source-Level Solutions

Section 3.5.5 shows that even our improved policy inevitably results in some over- and underestimation, which raises an interesting question whether it is possible to further improve the policies so that their accuracy approaches that of source-level solutions. Here we present three scenarios where a compiler makes the task of accurately

```
1  5a0d32:  xor    %esi,%esi
2  5a0d34:  xor    %edx,%edx
3  5a0d36:  mov    %rbp,%rdi
4  #struct ref *(*get_refs_list)(struct transport *transport, int for_push,
       const struct argv_array *ref_prefixes);
5  5a0d39:  callq  *0x10(%rax)
```

Listing 3.11 Immediate zero and NULL as arguments

recovering function signatures undecidable, and therefore show that binary-level techniques can never achieve the accuracy of source-based solutions.

3.6.2   *Immediate Value Zero vs. NULL Pointer*

A simple example demonstrating the limitation of binary analysis in this context is the differentiation between an immediate value zero and the NULL pointer. Line 4 of Listing 3.11 shows a callee function with the second and third arguments being integer and pointer type, respectively, while Line 1–2 show the caller preparation with identical instructions for these two arguments. It clearly demonstrates that binary analysis is unable to distinguish the two cases and would have to make approximations in recovering the caller signature.

3.6.3   *Arguments Unused*

Another scenario arises in the case of unused arguments at the callee (corresponding to complication case *Unread*), where binary analysis cannot differentiate

- Listing 3.12a: a callee function with an argument passed in but the argument is not used; and
- Listing 3.12b: a callee function without arguments.

Binary analysis would not be able to differentiate the two cases as observations on their parameter-passing registers are identical.

3.6.4   *Registers Overloading*

Registers are used for passing arguments as well as any other general purposes (corresponding to complication case *Temp*), and binary analysis usually cannot distinguish the two cases. Listing 3.13 shows two indirect callers with

```
1  bfd_plugin_core_file_failing_signal (bfd *abfd )
2  482000: push   %rax
3  482001: mov    $0x4dc9e1,%edi
4  482006: mov    $0x1ac,%esi
5  48200b: callq  405230 <bfd_assert>
```

a Argument passed in but not used

```
1  void bfd_section_already_linked_table_free ()
2  48aa60: mov    $0x7172f8,%edi
3  48aa65: jmpq   406860 <bfd_hash_table_free>
```

b No argument

Listing 3.12 Function argument unused

```
1  51e199:mov   %eax,%esi
2  51e19b: test  %r15,%r15
3  51e19e:je   51e1ad
4  51e1a0:lea   0xe0(%rsp),%rdi
5  #(fptr_T)(func_one(&cc, c));
6
7  51e1a8: callq *%r15
```

```
1  43ae62: mov   %ebp,%esi
2  43ae64: test  %rax,%rax
3  43ae67: je    43ae6f
4  43ae69: mov   %ebp,%edi
5  #get_elf_backend_data(abfd)->
      obj_attrs_order(i);
6  43ae6b: callq *%rax
```

a %esi used to pass argument   **b** %esi used to store temporary

Listing 3.13 Example of argument register usage

- Listing 3.13a: a caller that uses %esi to pass the second argument to callee.
- Listing 3.13b: a caller that uses %esi to store a temporary value.

Again, binary analysis would not be able to tell apart these two cases and an approximation has to be made in extracting function signatures.

We stress that this is not an exhaustive list of cases where binary analysis may fail, but the three scenarios identified are specific to funciton signature recovery where compiler optimization makes binary analysis *undecidable*.

3.6.5 Security Implication with Imprecise Function Signature Recovered

The undecidability in binary analysis results in inevitable errors in function signature recovery from (optimized) binary executables. An immediate question, therefore, is on the extent to which such errors impact security applications. In this subsection, we evaluate this security implication from two perspectives.

Imprecision on the Set of Callees Allowed. Our first evaluation focuses on the number of callees allowed in a CFI enforcement, and here we consider six solutions:

- AT [12]: A binary-level solution that allows indirect callers to target any "Address-Taken" functions;
- TypeArmor [19]: A binary-level solution with function signatures capturing the number of arguments;
- τCFI [18]: A binary-level solution with function signatures capturing the number of arguments and width of arguments;
- Our improved policy: A binary-level solution with function signatures capturing the number of arguments and width of arguments, targeting optimized binaries; and
- IFCC [9]: A (relatively old) source-level solution with function signatures capturing the number of arguments; in LLVM-3.4.
- LLVM-CFI[8]: A (latest) source-level solution with more precise function signatures (the number of arguments and their primitive types, function return type) captured; in LLVM-10.0.

Table 3.9 shows the median of the number of callees allowed for each indirect caller for the 1,344 applications in our test suite under different policies. We can see that compared to AT, TypeArmor, τCFI, and our improved policies reduce the number of legal control-transfer targets by about 20%, 49%, and 54%, respectively, while none of the binary-level solutions could achieve precision of source-level techniques. In particular, LLVM-CFI achieves much better accuracy because it uses finer-grained types of arguments—char* and const char*, struct A* and struct B* are considered different types—which cannot be differentiated at binary level.

Effectiveness in Allowing/Disallowing COOP Gadgets. With Table 3.9 showing the number of mistakes each solution makes, we next evaluate the extent to which these mistakes result in initial COOP gadgets an attacker could use to construct code-reuse attacks. This time, we only focus on τCFI and our improved policy as they run relatively close in the previous evaluation. We use the same heuristics proposed in the corresponding papers to find potential Main-Loop Gadgets (ML-G) [17] and RECursive Gadgets (REC-G) [30] for all C++ applications in our test suite. Table 3.10 shows the total number of such gadgets as well as the number of such gadgets whose function signatures are correctly identified by τCFI and our improved policy. Bigger numbers indicate better effectiveness of CFI in disallowing the corresponding code-reuse attacks.

As we can see, τCFI correctly identifies 68% and 64% ML- and REC- gadgets, respectively, while our improved policy achieves 78% and 74% effectiveness, respectively. We believe that this evaluation provides a good indicator on the security impact of our improved CFI policies.

---

[8]https://clang.llvm.org/docs/ControlFlowIntegrity.html.

Table 3.9 Number of callees allowed by different policies

		Opt	AT	TypeArmor	τCFI	Improved	IFCC	LLVM-CFI
clang	C	O0	543	412	290	246	114	7
		O1	540	446	242	213	124	8
		O2	394	318	147	147	93	7
		O3	380	300	130	120	99	8
	C++	O0	3,379	2,734	2,343	2,186	1052	37
		O1	3,290	2,631	1,879	1,805	998	35
		O2	702	552	304	270	251	44
		O3	710	543	296	284	247	44
gcc	C	O0	546	499	336	257		
		O1	446	373	272	239		
		O2	418	318	147	147		
		O3	406	332	231	200		
	C++	O0	4,505	3,920	3,278	3,219		
		O1	686	498	314	301		
		O2	698	477	294	281		
		O3	656	527	315	299		
Geomean			767	612	395	353	232	19

Table 3.10 Potential ML-G and REC-G gadgets

	Opt	ML-G			REC-G		
		icall	τCFI	Improved	icall	τCFI	Improved
clang	O0	93	53	64	73	41	45
	O1	58	50	50	56	44	44
	O2	70	46	52	60	41	44
	O3	70	42	53	49	35	39
gcc	O0	96	50	68	71	32	46
	O1	98	71	80	74	50	56
	O2	113	100	103	33	21	30
	O3	106	79	84	22	15	17
Geomean		83	56	65	58	37	43

Severity of Each Mistake. For each mistake in recovering function signature of the caller, we check how far the mistake is from the ground truth, which also has a direct implication on the amount of flexibility an attacker has when using the corresponding caller to construct an code-reuse attack. Figure 3.3 shows the result of this evaluation, again, on our test suite of 1,344 applications, with x-axis labels being:

- $+t$: the average number of indirect callers whose number of arguments is overestimated by t; and

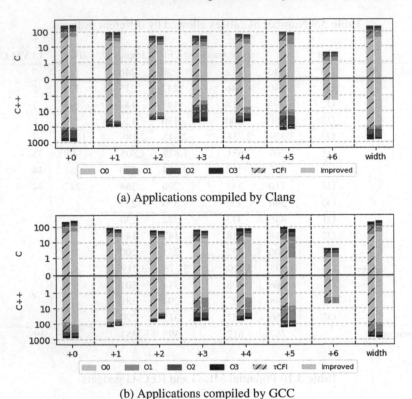

(a) Applications compiled by Clang

(b) Applications compiled by GCC

Fig. 3.3 Amount of flexibility of code-reuse attacks in each mistake in function signature recovery of indirect callers

- width: the average number of indirect callers whose function signature (number and width of arguments) is correctly recovered.

Besides showing the consistently better results from our improved policy compared to those from τCFI, we also notice that our improved policy performs most significantly better on "+5", which means our improved policies manage to correct a larger number of more severe mistakes made by τCFI.

3.7 Summary

In this chapter, we study how compiler optimization impacts function signature recovery implemented by TypeArmor and τCFI. Our study shows that compiler optimization has important impact on function signature recovery and potentially results in unmatched function signatures at callees and callers. In order to better deal with these optimizations, a set of improved policies is proposed, with results showing that most intricacies identified earlier being mitigated.

References

1. M. Abadi, M. Budiu, Ú. Erlingsson, J. Ligatti, Control-flow integrity principles, implementations, and applications. ACM Trans. Inf. Syst. Secur. 13(1), 4 (2009)
2. T. Bletsch, X. Jiang, V.W. Freeh, Z. Liang, Jump-oriented programming: a new class of code-reuse attack, in *Proceedings of the 6th ACM Symposium on Information, Computer and Communications Security* (ACM, 2011), pp. 30–40
3. S. Checkoway, L. Davi, A. Dmitrienko, A.-R. Sadeghi, H. Shacham, M. Winandy, Return-oriented programming without returns, in *Proceedings of the 17th ACM Conference on Computer and Communications Security* (ACM, 2010), pp. 559–572
4. Nergal. The advanced return-into-lib(c) exploits (2001), http://phrack.org/issues/58/4.html
5. H. Shacham, The geometry of innocent flesh on the bone: return-into-libc without function calls (on the x86), in *Proceedings of the 14th ACM Conference on Computer and Communications Security* (ACM, 2007), pp. 552–561
6. A.J. Mashtizadeh, A. Bittau, D. Boneh, D. Mazières, CCFI: cryptographically enforced control flow integrity, in *Proceedings of the 22nd ACM Conference on Computer and Communications Security* (ACM, 2015), pp. 941–951
7. B. Niu, G. Tan, Modular control-flow integrity, in *Proceedings of the 21st ACM Conference on Computer and Communications Security* (ACM, 2014), pp. 577–587
8. B. Niu, G. Tan, Per-input control-flow integrity, in *Proceedings of the 22nd ACM Conference on Computer and Communications Security* (ACM, 2015), pp. 914–926
9. C. Tice, T. Roeder, P. Collingbourne, S. Checkoway, Ú. Erlingsson, L. Lozano, G. Pike, Enforcing forward-edge control-flow integrity in {GCC} & {LLVM}. in *Proceedings of the 23rd USENIX Security Symposium* (2014), pp. 941–955
10. C. Lindig, Random testing of C calling conventions, in *Proceedings of the 6th International Symposium on Automated Analysis-Driven Debugging* (ACM, 2005), pp. 3–12
11. C. Zhang, T. Wei, Z. Chen, L. Duan, L. Szekeres, S. McCamant, D. Song, W. Zou, Practical control flow integrity and randomization for binary executables, in *Proceedings of the 34th IEEE Symposium on Security and Privacy* (IEEE, 2013), pp. 559–573
12. M. Zhang, R. Sekar, Control flow integrity for cots binaries, in *Proceedings of the 22nd USENIX Security Symposium* (2013), pp. 337–352
13. N. Carlini, A. Barresi, M. Payer, D. Wagner, T.R. Gross, Control-flow bending: on the effectiveness of control-flow integrity, in *Proceedings of the 24th USENIX Security Symposium* (2015), pp. 161–176
14. L. Davi, A.-R. Sadeghi, D. Lehmann, F. Monrose, Stitching the gadgets: on the ineffectiveness of coarse-grained control-flow integrity protection, in *Proceedings of the 23rd USENIX Security Symposium* (2014)
15. I. Evans, F. Long, U. Otgonbaatar, H. Shrobe, M. Rinard, H. Okhravi, S. Sidiroglou-Douskos, Control jujutsu: on the weaknesses of fine-grained control flow integrity, in *Proceedings of the 22nd ACM Conference on Computer and Communications Security* (ACM, 2015), pp. 901–913
16. E. Göktas, E. Athanasopoulos, H. Bos, G. Portokalidis, Out of control: overcoming control-flow integrity, in *Proceedings of the 35th IEEE Symposium on Security and Privacy* (IEEE, 2014), pp. 575–589
17. F. Schuster, T. Tendyck, C. Liebchen, L. Davi, A.-R. Sadeghi, T. Holz, Counterfeit object-oriented programming: on the difficulty of preventing code reuse attacks in C++ applications, in *Proceedings of the 36th IEEE Symposium on Security and Privacy* (IEEE, 2015), pp. 745–762

18. P. Muntean, M. Fischer, G. Tan, Z. Lin, J. Grossklags, C. Eckert, τCFI: type-assisted control flow integrity for x86-64 binaries, in *Proceedings of the 21st International Symposium on Research in Attacks, Intrusions, and Defenses* (Springer, 2018), pp. 423–444
19. V. Van Der Veen, E. Göktas, M. Contag, A. Pawoloski, X. Chen, S. Rawat, H. Bos, T. Holz, E. Athanasopoulos, C. Giuffrida, A tough call: mitigating advanced code-reuse attacks at the binary level, in *Proceedings of the 37th IEEE Symposium on Security and Privacy* (IEEE, 2016), pp. 934–953
20. C. Kruegel, W. Robertson, F. Valeur, G. Vigna, Static disassembly of obfuscated binaries, in *Proceedings of the 13th USENIX Security Symposium* (USENIX Association, 2004)
21. Ghidra. The ghidra decompiler (2019), https://ghidra-sre.org/
22. M. Matz, J. Hubicka, A. Jaeger, M. Mitchell, System V application binary interface, *AMD64 Architecture Processor Supplement, Draft v0.99* (2014)
23. Intel® 64 and IA-32 architectures software developer's manual (2018)
24. J. Lee, Y. Kim, Y. Song, C.-K. Hur, S. Das, D. Majnemer, J. Regehr, N.P. Lopes, Taming undefined behavior in LLVM, in *Proceedings of the 38th ACM SIGPLAN Conference on Programming Language Design and Implementation* (ACM, 2017), pp. 633–647
25. A. Milburn, H. Bos, C. Giuffrida, Safelnit: comprehensive and practical mitigation of uninitialized read vulnerabilities, in *Proceedings of the 24th Network and Distributed System Security Symposium* (2017), pp. 1–15
26. C. Lattner, V. Adve, LLVM: a compilation framework for lifelong program analysis & transformation, in *Proceedings of the 2nd International Symposium on Code Generation and Optimization* (IEEE Computer Society, 2004)
27. DWARF Debugging Information Format Committee, et al.: DWARF debugging information format, version 4. Free Standards Group (2010)
28. N. Rosenblum, B.P. Miller, X. Zhu, Recovering the toolchain provenance of binary code, in *Proceedings of the 20th International Symposium on Software Testing and Analysis* (ACM, 2011), pp. 100–110
29. N.E. Rosenblum, B.P. Miller, X. Zhu, Extracting compiler provenance from program binaries, in *Proceedings of the 9th ACM SIGPLAN-SIGSOFT Workshop on Program Analysis for Software Tools and Engineering* (ACM, 2010), pp. 21–28
30. S.J. Crane, S. Volckaert, F. Schuster, C. Liebchen, P. Larsen, L. Davi, A.-R. Sadeghi, T. Holz, B. De Sutter, M. Franz, It's a trap: table randomization and protection against function-reuse attacks, in *Proceedings of the 22nd ACM SIGSAC Conference on Computer and Communications Security* (2015), pp. 243–255

Chapter 4
Control-Flow Carrying Code

4.1 Introduction

In the previous chapter, we introduce the approach to generate a more accurate CFG by making use of function signature matching, in this chapter, we will show how to implement the CFI policy securely.

An assumption made in most existing CFI approaches, including coarse-grained [1, 2] and fine-grained [3–5] ones, is that read-only data and code sections cannot be overwritten by attackers. For example, CFI proposed by Abadi et al. [6] relies on read-only tags inside the code segment, and numerous approaches use a table structure (made read-only) to store valid targets of indirect branches [1, 3, 4]. However, there are scenarios in which such page-level protection is unavailable, e.g., bare-metal systems which do not have a Memory Management Unit (MMU) and applications with dynamically generated code. Moreover, data race attacks [7], Rowhammer attacks [8] and Data-Oriented Programming (DOP) [9] have demonstrated that it is possible to gain arbitrary memory read and write access.

In this chapter, we explore the possibility of enforcing CFI in the absence of such an assumption. Specifically, we look into encoding CFI policies into the machine instructions directly without relying on policies specified in additional data structures (i.e., the read-only table structures in existing CFI approaches) or inserting CFI checks into the code segment. The general idea is to embed a statically constructed CFG to the instructions, execution of which is conditioned on correct control flows. In this way, each intended instruction will carry a proof that can validate the control-flow transfer. Unintended instructions cannot be executed as the proof in these instructions are not correct. Intuitively, instructions with CFG embedded can be seen as a proof-carrying code [10], where this proof is self-contained in the code rather than being encoded into a separate table. The challenge is how to embed the CFG into the instructions and how to correctly execute them at runtime.

Inspired by the framework of Instruction-Set Randomization (ISR) [11] where instructions of a program are encrypted with a secret key, we present Control-Flow Carrying Code, C^3, which encrypts each basic block in the program with a key

© The Author(s), under exclusive license to Springer Nature Switzerland AG 2021   53
Y. Lin, *Novel Techniques in Recovering, Embedding, and Enforcing Policies for Control-Flow Integrity*, Information Security and Cryptography,
https://doi.org/10.1007/978-3-030-73141-0_4

derived from the CFG. More specifically, the key is derived from (the addresses of) valid callers of the basic block to ensure correct control-flow transfers. At runtime, only the valid callers (their addresses) could enable the correct reconstruction of the key to decrypt the basic block. In this way, C^3 manages to embed and enforce CFI in the program instructions.

However, two challenges remain in making C^3 practical. First, a basic block may have multiple valid callers. These valid callers have different addresses, while the successor block has to be encrypted with a single key. How does C^3 enable the reconstruction of the single correct key by all the valid control-flow transfers? To address this challenge, C^3 utilizes the secret sharing scheme [12] to make the key shared among valid callers.

Although secret sharing helps solve this important challenge at a high level, we encounter more challenges in its application in our setting. For example, secret sharing requires that *all (a variable number of)* callers of the basic block be on the same secret sharing curve. The implication is that once we have the curve fixed, addresses of these callers can no longer take arbitrary locations but have to be on the secret sharing curve determined. This imposes extra challenges in laying basic blocks in the text segment of the program. To address this, we design an algorithm to redistribute basic blocks to positions satisfying the secret sharing curve.

We have implemented C^3 that consists of two components, one that performs binary rewriting to redistribute and encrypt basic blocks, and the other as a plug-in to an existing instrumentation platform to assist runtime execution of the rewritten executable. We apply C^3 to a number of server and non-server applications on the Linux platform. Our experimental results demonstrate that C^3 effectively defends against control-flow hijacking attacks and at the same time, introduces realistic runtime performance overhead for server applications comparable to existing Instruction-Set Randomization (ISR) implementations on the same instrumentation platform. Similar to the arguments in ISR systems, we believe that such overhead could be significantly reduced with a hardware-assisted platform.

4.2   Overview of C^3

4.2.1   Threat Model and Assumptions

The proposed defense, C^3, is aimed to protect a vulnerable application against control-flow hijacking attacks such as ROP attacks. The application to be protected may have some vulnerabilities that can be leveraged by an attacker to inject an exploit payload (code or data). We focus on user-space attacks leaving kernel exploits out of our scope. Specifically, we assume that:

• The target program does not contain self-modifying or dynamically-generated code.

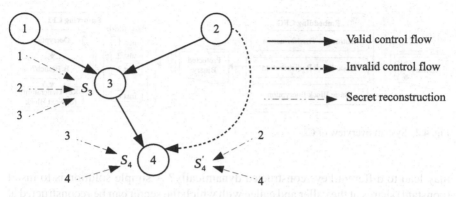

Fig. 4.1 Example of secret reconstruction

- Attackers could use attacks to bypass W⊕X, such as Data-Oriented Programming [9], data race [7] and Rowhammer attacks [8], and could exploit information disclosure vulnerabilities to investigate the victim's process memory.
- Since the current implementation of C³ is on top of the popular instrumentation platform Pin, we assume that attackers do not target Pin in their attacks and the partial memory segment managed by Pin (e.g., the code cache) is secure. This assumption can be removed if C³ is supported by native hardware.

4.2.2 Embedding CFG to Instructions

Rather than consulting additional information stored in read-only memory, we propose to embed CFG to instructions. An instruction with CFG embedded can check the integrity of the control flow automatically during the execution without querying other data structures. In particular, C³ embeds the CFG information by encrypting each basic block (an idea inspired by ISR [11]) with a key generated from control-flow dependent information. At runtime, the basic blocks are decrypted using a key reconstructed from the actual control-flow transfers taken. Only when the correct control flow paths are taken will the instructions be decrypted correctly.

In Fig. 4.1, each node represents an encrypted basic block while edges indicate control flows. The solid edges represent valid control flows with S_i indicating the encryption key for basic block i. S_3 and S_4 are generated according to the valid control-flow path $<1, 3>$, $<2, 3>$ and $<3, 4>$. When there is an invalid control-flow transfer from node 2 to 4 denoted by the dotted edge, a wrong key S_4' is constructed which would result in illegal instruction faults.

Although the idea sounds straightforward, there are multiple design questions and challenges. First, what information do we use to generate the key? Such information needs to be both statically and dynamically available, and it shall allow enforcement of CFI. How do we deal with basic blocks involved in multiple control flows, which

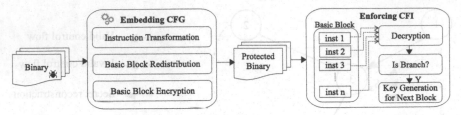

Fig. 4.2 System overview of C^3

may lead to different keys constructed dynamically? A simple solution is to insert
(constant) shares at the caller and callee with which the secret can be reconstructed at
runtime. However, such an approach does not provide Control-flow integrity because
an attacker can reuse the share at other caller sites.

Our solution is to use the addresses of the branch transfer instruction and its target
as the shares since they capture the control transfer information precisely. To deal with
basic blocks involved in multiple control flows, we use basic block redistribution and
secret sharing [12] to encode the key. Figure 4.2 shows an overview of C^3, consisting
of two components.

- **Embedding CFG.** C^3 transforms branch transfer instructions (indirect branches,
  conditional jumps, and direct calls) to have a secret share embedded, and then
  redistributes basic blocks to specific addresses so that all valid callers are on the
  same secret sharing curve. Finally, basic blocks are encrypted with the secret.
- **Enforcing CFI.** Whenever the program attempts a control transfer, C^3 obtains the
  caller and callee addresses and reconstructs the key to decrypt the callee basic
  block before control transfer takes place.

4.3 Detailed Design of C^3

C^3 takes as input a binary executable (without source code) and outputs a modified
executable with CFG embedded and CFI enforced.

4.3.1 Secret Sharing and Challenges

As discussed in Sect. 4.2, our approach of embedding CFG into instructions is to
encrypt a basic block and to enable decryption with any correct control transfer.
For a basic block with multiple callers, we can imagine that every valid caller shall
contribute to the encryption key; however, in a concrete execution, only one valid
caller is involved and the decryption key is reconstructed. This is where the idea
of secret sharing comes to our design—only part of the ingredients of the secret

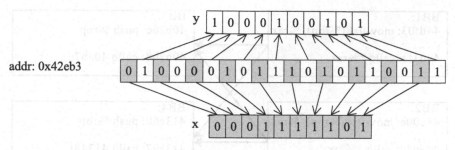

Fig. 4.3 Extracting x and y for address 0x42eb3

key is needed for correct reconstruction. C³ uses Shamir's approach [12] due to its simplicity.

The next question is the degree of the secret sharing equation. A general guideline is to keep it small to minimize overhead. We can use a degree of two with the source and target addresses of the control transfer—the minimum information to fully describe a control transfer. However, this runs into the risk of a code pointer disclosure exploit that discloses both addresses and allows an attack to decrypt the basic block. To counter such an attack, we add one random value (called the master key) which is unknown to the attacker to construct the secret key. Specifically, we use a degree of three, with the secret sharing equation

$$y = a_0 + a_1x + a_2x^2 \ (mod \ M) \tag{4.1}$$

where a_0 is the secret key for encryption and decryption, and x, y are k-bit coordinates extracted from the source and target addresses and the master key. C³ obtains x and y from the lower-order odd- and even-index bits of an address (see Fig. 4.3 for an example). Reconstruction of the secret follows Eq. 4.2 with $x = 0$.

$$y = \sum_{i=1}^{3} y_i \prod_{1 \le j \le 3, j \ne i} (x - x_j)(x_i - x_j)^{-1} \ (mod \ M) \tag{4.2}$$

To support a basic block with multiple callers, we can simply relocate the caller instructions so that they all lie on the parabola. However, a more challenging issue is to support a set of basic blocks with the same (set of) callers. Figure 4.4 shows an example with $BB3$ and $BB4$ having the same set of callers $BB1$ and $BB2$. Following the secret sharing design we outline above, the two parabolas for $BB3$ and $BB4$ will have three intersection points—the master key, $BB1$, and $BB2$; however, different parabolas could have up to two intersections only. Therefore, C³ not only needs to relocate the basic blocks to move them onto specific parabolas, but also needs to perform some special transformations to control transfer instructions; see the next subsection.

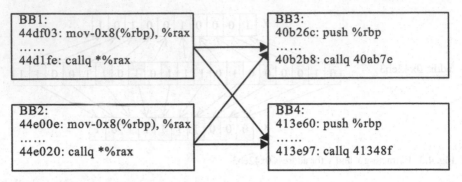

Fig. 4.4 Multiple callers to multiple callees

4.3.2   Instruction Transformation

In fact, the challenge shown in Fig. 4.4 is not the only one that C^3 needs to handle.

- **C1: Multiple callers to multiple callees.** In such cases, secret sharing curves for the callees have three or more intersections (including the master key), which is not possible for parabolas. We add an intermediate block between the callers and callees so that multiple callees now have a single caller.
- **C2: Basic blocks that are not freely movable.** Examples of such blocks include targets of `ret` instructions which must follow the `call` instruction, and the default branch of conditional instructions which must follow the conditional branch instruction. They cannot be moved freely to other locations due to the implicit control flow. Our strategy is to transform the implicit control flows into explicit ones.
- **C3: Basic blocks with multiple entries.** Multiple entries will lead to different keys derived for the same basic block. Our strategy is to break it up into multiple basic blocks, each of which has a single entry.

In the rest of this subsection, we use an example (Fig. 4.5) to explain how C^3 solves these complexities. Note that the transformation is via binary rewriting without source code of the program.

Transforming Indirect Call and Indirect Jump Instructions
C^3 transforms an indirect call instruction into two `push` instructions (one to save the return address and the other to save the target address onto the stack) followed by a `jmp` instruction (jumping to a common stub); see $BB5$ and $BB5'$ in Fig. 4.5. The stub block has a single `ret` instruction.

Although this simple transformation solves C1, it potentially enforces a relaxed CFI policy since multiple control transfer targets now go through the same common stub block. We stress that the same policy is used by existing coarse-grained CFI methods [1, 2]. Moreover, C^3 increases the difficulty of a stealthy attack since the valid targets are now encrypted. We could use a more complicated secret sharing curve to enforce a finer-grained policy, but C^3 chooses this solution due to its simplicity and

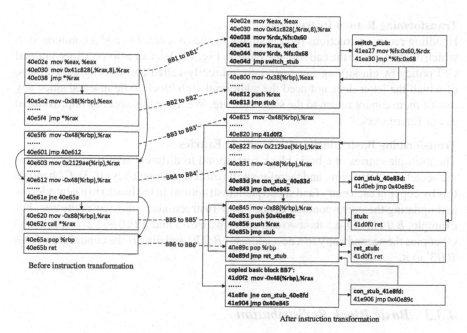

Fig. 4.5 An example of instruction transformation by C³

enforcing a CFI policy not less secure than existing work. Note that a byproduct of pushing the return address on the stack (the first push in $BB5'$) is a solution to C2, as the return site can now be freely moved (explained later in the next subsection).

Indirect jumps are handled in the same way, except that we only need one push instruction since there is not a return address, e.g., $BB2$ in Fig. 4.5. Additional challenge arises here when the indirect jump was generated due to switch/case statements during compilation, where local variables are sometimes accessed via %rbp directly without changing %rsp. In such cases, we cannot simply push the target address of the indirect jump onto the stack because doing so would overwrite the local variables. Instead, we make use of thread local storage to store the target; see the indirect jump in $BB1$ of Fig. 4.5. In order to transform an indirect jump jmp *0x8(%rax) (the target is the address in memory) while having the same switch stub with jmp *(%rax), we simply move the target of them to the temporary register %rdx as shown in $BB1'$.

Transforming Conditional Jump Instructions
Conditional jumps usually have a fall-through branch to the instruction that immediately follows, forming an implicit control transfer (C2). We turn this into an explicit one by inserting a direct jump instruction as in $BB4'$ of Fig. 4.5. Note that similar to indirect jumps, conditional jumps may be followed by multiple callees (C1); that is why we also add a stub block as shown in $BB4'$ of Fig. 4.5.

Transforming Return Instructions

Handling return instructions (C1) is simple as we only need to add a common stub which then returns to the call site; see $BB6'$ in Fig. 4.5. We can enforce a finer-grained CFI policy by classifying functions into indirectly-called and directly-called ones, of which the latter does not need the additional stub block to be inserted since any two of them cannot return to the same call site. We leave this security improvement as our future work.

Transforming Basic Blocks with Multiple Entries

The multiple entries of a basic block correspond to different sets of ingredients for the secret reconstruction, and therefore will result in different keys (C3). C^3 handles this by copying each entry (and subsequent instructions in the block) to a new address and updating the corresponding control-flow instructions to the new addresses. For example, $BB4$ in Fig. 4.5 has two entries, 0x40e603 and 0x40e612, respectively. C^3 copies the second entry to a new address ($BB7'$) and directs the control flow from $BB3'$ to it.

4.3.3 Basic Block Redistribution

Redistributing basic blocks so that all callers of a control transfer are on the same secret sharing curve is an interesting and non-trivial problem. One can consider it as a directed graph traversal in which whenever a node is traversed, we pick a parabola and ensure that all its callers are on it by moving some or all the callers. However, if the traversal is not carefully designed, we could get into a failure where a node that has been previously moved on a parabola now needs to be moved again to satisfy another parabola—a mission impossible. Therefore, the key is to design a directed graph traversal algorithm that minimizes or eliminates such a risk.

C^3 uses a customized Depth First Search (DFS) algorithm. Intuitively, DFS fits our requirement in that it explores a branch to its ultimate leaves before backtracking or stepping into a new branch, which avoids unnecessary moving of caller nodes of branches already unexplored. We customize it with a "look ahead" capability which switches to another nearby branch when continuing exploring the current branch will get into a "mission impossible" case.

As shown in Fig. 4.6 where shaded nodes denote those that had previously been moved (and therefore cannot be moved again) and hollow ones otherwise, continuing to traverse node A would run into a failure mode since node B will have two caller nodes fixed, making it impossible to find a parabola for node B (it already has three points determined including the master key). In this scenario, our "look ahead" function will traverse the sub-branch of node B before going back to traverse node A. This "look ahead" function is also used to decide the starting point. By default, C^3 picks a node with the largest number of callers as the starting point, and then uses the "look ahead" function to check whether this starting point and one of its callers target the same basic block. If they do, C^3 uses this basic block as the starting point. The detailed algorithm is shown in Algorithm 1.

Fig. 4.6 "Look ahead" DFS search

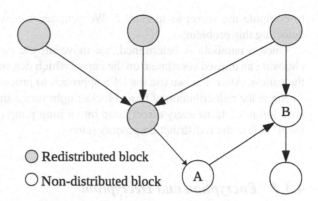

○ Redistributed block

○ Non-distributed block

Algorithm 1. Basic Block Redistribution

1: **procedure** REDISTRIBUTION(callee, master_key, k, p)
2:   **if** callee not in key_block **then**
3:       *priority_callee = Look_Ahead(callee)*
4:       **if** priority_callee **then**
5:           Redistribution(*priority_callee, master_key, k, p*)
6:       *callers* = callee_caller[*callee*]
7:       *moved_callers* = find_moved_callers(*callers*)
8:       **if** len(moved_callers) == 0 **then**
9:           *caller* = random_choose_caller(*callers*)
10:      **if** len(moved_callers) == 1 **then**
11:          *caller* = moved_callers[0]
12:      compute_key(*callee, caller, master_key, k, p*)
        ▷ % move all callers of this basic block to be on the curve %
13:      **for** i in callers **do**
14:          **if** i not in redistributed_block **then**
15:              move_caller(*callee, i, master_key, k, p*)
        ▷ % DFS: process callees of this basic block %
16:      **for** i in caller_callee[callee] **do**
17:          Redistribution(*i, master_key, k, p*)
        ▷ % backtracking %
18:      **for** i in callers **do**
19:          **for** j in caller_callee[i] **do**
20:              Redistribution(*j, master_key, k, p*)

Specifically, for a callee to be processed, we first check whether there is a prior basic block using the "look ahead" mechanism described above. Then, for each callee to be processed, we check whether there exists any of its callers that has a fixed address. If there is, we use this caller (with a fixed address) to determine the parabola; otherwise we randomly choose a caller to determine the parabola. The special and additional processing here is that for each (caller or callee) address, we need to check whether it will have the same x value with its callee, caller or the master key, since the same x value could result in a failure in calculating the inverse to compute the secret as in Eq. 4.2. We generate a new random address if when detecting this problem.

Once a parabola is determined, we move all the callers onto it by randomly choosing an unused coordinate on the curve, which determines the new addresses of the callers. After that, we use the DFS approach to process other basic blocks.

Since the redistribution of basic blocks might turn a short jump instruction into a long jump, C^3 turns every direct jump into a long jump (with a four-byte displacement) before the redistribution process starts.

4.3.4 Encryption and Decryption

Before we present details of C^3 in encrypting a basic block, we note that completely separating code from data into different sections is a prerequisite for our encryption to work. This is because the encryption of any data may disrupt program execution when it is not decrypted at runtime. Fortunately, many linkers are configured to ensure such separation, and compiler optimizations like jump tables are also typically moved to a non-code section. C^3 does not include PLT calls in its protection as doing so will result in .plt section containing non-continuous addresses due to basic block redistribution (see the previous subsection), which in turn makes it impossible for the dynamic loader to update addresses in the Global Offset Table (GOT).

C^3 uses XOR as the encryption function due to its simplicity. The reconstructed secret s from secret sharing is used as the seed to a pseudo-random function generator to generate a 16-bit key for encryption. The length of the secret s is a configurable parameter which has an upper bound of 16 because going beyond that may result in distance between two instructions greater than 2^{31}. To fight against memory disclosure attacks that attempt to compromise the master key, C^3 stores the master secret key outside of the binary into a database file, an approach used in some ISR approaches [13]. We note that C^3 could also perform load-time encryption on the basic blocks using a session key (replacing the master key) to further improve security [14, 15]. Also note that when the binary rewriting process is performed remotely, we could make use of remote attestation [16] to securely distribute the master key. We leave both ideas as our further work.

4.3.5 Transitioning from Unprotected to Protected Code

C^3 supports partial protection of a program that contains protected (CFG embedded) and unprotected (e.g., system or third-party libraries without CFG embedded) code. However, the transitioning from unprotected to protected code needs special attention since CFI checks will fail as the caller is not on the secret sharing curve of the callee. Such transitioning typically occurs in two scenarios.

- **Returning to protected code.** This happens when protected code calls an external library function and subsequently returns from it.
- **Calling to a function in protected code.** This happens when the external library function (e.g., qsort, bsearch) calls a comparison function in the protected code.

We handle these cases by adding a dummy block before each return target and function entry in the protected code, since we cannot accurately identify calls to a library function and functions called by the library. This dummy block has only one instruction that jumps to the actual target, and is encrypted with a key generated from its address. C^3 transfers control to the dummy block when detecting a control transfer from unprotected to protected code, the range of which is recorded into a (secure) database.

In this way, C^3 ensures that these dummy blocks cannot be invoked by control-flow transfers in the protected code and provides the same level of protection compared with existing CFI techniques.

4.4 Implementation

We implemented C^3 on an Ubuntu 64-bit system supporting inputs of ELF binary executables without source code.

4.4.1 Binary Rewriter

We developed our custom binary rewriter in 6,500 lines of Python code with the help of the disassembly engine Capstone [17]. The binary rewriter takes as input the ELF executable to be protected and the configuration of k. Embedding CFGs to an executable consists of three stages.

Before we embed control-flow information, we first obtain the static CFG. We do this by modifying a recent work typearmor [18] (which builds on Dyninst [19]).

Secondly, we use Capstone to disassemble the binary. C^3 uses the algorithm described in Algorithm 1 to select basic blocks and then compute the secret for each of them. Note that the redistribution algorithm will likely distribute basic blocks apart from each other, and many NOP instructions need to be inserted into the .text section.

In the last stage, we update the corresponding section information including program entry point, program header, section header, items in relocation table, .dynamic, and .dynsym sections. In addition, some instructions need to be updated to maintain the original control flow:

- **Direct jumps:** We transform all indirect branch transfers to jump to the stub first; see Sect. 4.3.2. Therefore, there are only direct jumps in the .text section now. The target address of a direct jump is specified as a relative offset from the address of the jump instruction, which needs to be recomputed after basic block redistribution.

- **PC-relative addressing mode:** We also need to patch instructions with PC-relative addressing mode, which are often used to generate position-independent code. The new x86 64 architecture natively supports PC-relative addressing, e.g., `lea ox200000 (%rip), %rbp` adds `0x200000` to the program counter and saves it to `%rbp`. To ensure correctness, C^3 updates these instructions by recomputing the new offset using the new program counter and the address of the redistributed target.
- **Function pointers:** They are usually absolute addresses of indirect call targets that are loaded into registers. To fix these instructions, the absolute address of the callee should be patched at the instruction that loads its address into the corresponding register. This is done by identifying all possible function pointers with the help of the symbol table and patching them to the redistributed addresses. The same goes to global function pointers where C^3 updates the address in data section.
- **Data pointers:** They need to be patched, too, because the starting offset of the data section has changed. C^3 patches them by adding the new offset to the original value.
- **Jump tables/virtual tables:** C^3 updates the base address of the jump table by adding the new offset to it. Patching virtual tables follows the same mechanism.

4.4.2 Execution Environment

We make use of Pin [20] to implement the execution environment with 1,100 lines of code in C++. It first reads from the secure database the master key and the protection range and then installs a callback that intercepts the loading of all images to obtain ranges of the unprotected memory. In addition to Pin, we can also make use of the dynamic code optimization platform called DynamoRIO [21] to implement the execution environment.

We then use the instrumentation callback at instruction granularity to detect a branch and compute the key for the next basic block. The decryption of basic blocks is performed by installing a callback that replaces Pin's default mechanism of fetching code from the target process. If the instruction fetched is within the range of protected code, we reconstruct the key from secret sharing parabola for decryption.

For code transitioning described in Sect. 4.3.5, we make use of PIN_Set ContextReg to set the value of `%rip` register to the address of the dummy block which has just one instruction that jumps to the actual target, and then use the PIN_ExecuteAt API to direct execution to it. For the transition from protected code to unprotected code, C^3 stops the decryption and let the code execute as normal. Similar to other CFI approaches, the attacker can use gadgets in unprotected code to construct code-reuse attack, which C^3 cannot defend against.

Table 4.1 Comparison with existing CFI techniques

Exploits	BinCFI [1]	CCFIR [2]	IFCC [5]	kBouncer [25]	ROPecker [26]	C^3
Göktas et al. [22]	✓	✓				
Davi et al. [23]	✓			✓	✓	
Conti et al. [24]	✓	✓	✓			
Hu et al. [9]	✓	✓	✓	✓	✓	

To avoid performing frequent key reconstruction for direct branch transfer instructions, we cache the key for subsequent use. Therefore, each direct branch transfer instruction corresponds to only one key reconstruction.

4.5 Evaluation

We first analyze the security of C^3 and then measure its performance overhead with real-world applications.

4.5.1 Security

C^3 mitigates code-injection attacks in the same way Instruction-Set Randomization defeats them—when control flow is redirected to injected code, C^3 will decrypt it into random bytes. The attacker could not prepare the correct encrypted code since she does not know the master key.

C^3 also mitigates most Code-Reuse Attacks (CRA) due to three reasons. First, C^3 generates a wrong key when an invalid control transfer happens, which results in a random byte stream to be executed. Second, redistributing and encrypting basic blocks makes it harder for attackers to analyze and locate gadgets, which defeats most static CRA. Finally, the encrypted basic blocks result in little information revealed even when an attacker manages to dump the execution memory, which defeats most dynamic CRA.

Comparison with Existing CFI Techniques

A number of recent proof-of-concept exploits have shown how existing coarse-grained CFI techniques can be bypassed [22–24]. Although C^3 also enforces a coarse-grained policy, its unique handling of basic blocks (encryption) provides a new defense to make these exploits unsuccessful. Table 4.1 compares various CFI techniques with C^3 on the CFI policy enforced and defense capability against the exploits.

As shown in Table 4.1, existing instrumentation-based CFI methods [1, 2, 5] do not insert checks for unintended control-flow transfers, making them vulnerable to the exploit proposed by Conti et al. [24]. Such an exploit would not work on C^3 as all instructions (intended or unintended) are encrypted. The exploit proposed by Hu et al. [9] succeeds on all existing CFI methods as they rely on the assumption that $W \oplus X$ is effective. Moreover, the content in the CFI table inserted by BinCFI provides sufficient information about useful gadgets if there is memory disclosure. However, since C^3 does not have this problem because it does not insert any metadata. The first three CFI approaches in Table 4.1 also suffer from TOCTOU attack—time of checking values of esp/rsp and time of executing ret, when the return address is stored in memory which could be modified by another thread. Under the protection of C^3, even if the address is modified by another thread, control flow will transfer to cipher-text which will result in program crashing.

Exploits that use call-preceded gadgets [22, 23] cannot succeed on C^3 since basic blocks are redistributed to random addresses. We performed experiments to verify the effectiveness of C^3 on defending against CRA that uses call-preceded gadgets using the test application ndh_rop from ROPgadget[1], a publicly available test set for ROP attacks. Our experiments verified that the payload that successfully exploits ndh_rop failed to run on C^3. Upon further investigation, we realized that it generated an illegal instruction fault when the return instruction directs control flow to the first call-preceded gadget. This is because this address is an invalid instruction which does not carry a valid proof to reconstruct the correct decryption key.

Compared with fine-grained approaches, e.g., Lockdown [27], which uses binary instrumentation to enforce CFI for different modules, C^3 can achieve better security as the attacker cannot make use of memory disclosure to traverse the memory of the victim program due to encryption of instructions. Basic block redistribution performed in C^3 can also be seen as effectively making the coarse-grained CFI policy finer-grained since the attacker cannot find the addresses of gadgets.

One may argue that the attacker could dump the protected code and do offline analysis to decrypt it. However, even if the attacker dumps the protected code and obtains the master key and the address of a basic block, she still has to try all possible encryption keys to see whether the basic block can be decrypted into valid instructions. We performed such experiments and realized that there are usually multiple such keys which have to be further tested on the resulting caller blocks for validity checks, and such checks have to carry on for callers of the callers, which makes it difficult for offline analysis to decrypt the protected code.

CFI Effectiveness with AIR

Zhang and Sekar [1] propose using *Average Indirect target Reduction (AIR)* for measuring the strength of CFI, which has become a common method of evaluation [27–29]. It computes the average number of machine code instructions that are eliminated as possible targets of indirect control transfers.

The formula used by Zhang and Sekar is shown in Eq. 4.3, where n is the number of indirect branch instructions in the program, and S is the total number of instructions to which an indirect branch can transfer control flow, whose value is the same as

---

[1]https://github.com/JonathanSalwan/ROPgadget.

Table 4.2 Average indirect target reduction

Programs	# of valid targets		AIR
	Ends with indirect branch	Ends with direct branch	
vsftpd ($k = 9$)	25	0	99.84%
Pure-FTPd ($k = 10$)	172	0	98.95%
ProFTPD ($k = 11$)	506	0	99.23%
httpd ($k = 11$)	171	0	99.74%
Nginx ($k = 11$)	125	0	99.81%
lighttpd ($k = 10$)	35	0	99.79%
Memcached ($k = 10$)	62	0	99.62%
Average			99.57%

the size of code in a binary. $|T_j|$ is the possible number of targets to which indirect branch j can transfer control flow after CFI enforcement.

$$\frac{1}{n}\sum_{j=1}^{n}(1 - \frac{|T_j|}{S}) \qquad (AIR) \qquad (4.3)$$

In the case of C^3, $|T_j|$ is the possible number of targets that can be interpreted as valid basic blocks for indirect branch j. Since we substantially increase the size of the .text section, instead of enumerating all possible addresses (which requires testing millions of addresses), we randomly choose 16, 384 addresses for effective testing when $k \leq 10$ and 65, 536 addresses for other k values. We consider all basic blocks ending with indirect transfer instructions as valid, and those ending with direct transfer instructions valid if their targets are in the .text section.

The results are shown in Table 4.2 with server applications. Interestingly, there are few addresses that can be interpreted as valid basic blocks, and all of them end with indirect transfer instructions. This is because C^3 extends the displacement in direct branches to four bytes, which makes the probability that a random sequence be interpreted as a valid direct branch small. On average, C^3 achieves an AIR value of 99.57%, comparable to existing CFI approaches [1, 27].

JIT-ROP
JIT-ROP [30] is an attack against fine-grained randomization. It assembles ROP gadgets "on-demand" without knowing the memory layout by exploiting the disclosure of a single code pointer. Specifically, the adversary traverses the memory space that the leaked pointer points to, searches for gadgets and cross-page transfer instructions to find new code pages and other useful gadgets. However, under C^3, a read performed from a code page yields cipher-text, which the adversary cannot disassemble without knowing the decryption key. As such, an adversary cannot use JIT-ROP to disclose new code pages to find gadgets.

To verify our intuition, we tried to use the ROP gadget finding tool peda[2] to identify gadgets in the protected binary nginx-1.4.0 after the loading phase, simulating the full disclosure of the code segment. Many gadgets ending with ret are found, which are chained together to form an attack payload. However, the gadgets found were based on encrypted basic blocks, which become invalid instructions and lead the execution into an illegal instruction fault.

Blind ROP

Blind ROP [31] uses the response from the victim process (crash vs. no crash) as a side channel to incrementally guess the position of a gadget. It assumes that the adversary can disassemble the code pages to find the required gadgets. Since the code pages are encrypted with C^3, Blind ROP will not succeed. We applied the exploit script provided by Bittau et al.[3] to nginx-1.4.0 protected by C^3, and found that it made all worker threads "stuck" as they were all running into an infinite loop of locating gadgets. Blind ROP uses a conservative implementation to incrementally populate the stack to find a stack-based stop gadget to avoid hanging. However, with C^3, every attempt in transferring control to this stack-based stop gadget results in a failure due to incorrect decryption of the callee block.

Control-Flow Bending

Control-Flow Bending (CFB) [32] bypasses conventional CFI that statically generates CFGs. CFB abuses certain functions whose executions may change their own return addresses to point to any call-preceded site which allows the attacker to "bend" the control flow. C^3 mitigates CFB attacks by preventing the attacker from locating call-preceded basic blocks—thanks to redistributing and encrypting of basic blocks.

Although C^3 successfully defends against these existing advanced control-flow hijacking attacks, we acknowledge that it is not necessarily effective against an attack specifically crafted for C^3. We further discuss this possibility in Sect. 4.6.

4.5.2   Performance Overhead

We evaluated C^3 with three FTP servers (vsftpd, ProFTPD, and Pure-FTPd), three web servers (Nginx, lighttpd, and Apache), a distributed memory caching system (Memcached), and some common applications (image processing tools sam2p, GraphicsMagic, and ImageMagics and bzip2). All programs are executed with their default settings on a desktop computer with an Intel i7-4510u CPU with 8 GB of memory running x64 version of Ubuntu.

To benchmark web servers, we configured Apache Benchmark[4] to issue 2,000 requests with 100 concurrent connections. For FTP servers, we configured

---

[2]https://github.com/longld/peda.
[3]http://www.scs.stanford.edu/brop/.
[4]http://httpd.apache.org/docs/2.4/programs/ab.html.

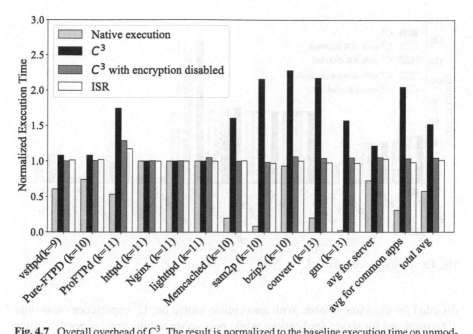

Fig. 4.7 Overall overhead of C^3. The result is normalized to the baseline execution time on unmodified Pin

pyftpbench benchmark[5] to open 20 connections and request 100 files per connection with over 100 MB of files requested. To benchmark Memcached, we used memslap.[6] We ran each experiment 10 times, ensuring that the CPUs were fully loaded throughout the tests, and report the median.

Since C^3 is implemented on top of the dynamic instrumentation platform Pin, we measure the performance of C^3 in terms of the additional execution overhead compared to these programs executing on an unmodified Pin v.3.5. To enable a better understanding of the results, we also report the execution overhead of another system that is built on top of Pin, namely Instruction-Set Randomization implemented by Portokalidis et al. [13].

Execution Time

We report, in Fig. 4.7, the execution time of each program under four settings: native execution, ISR [13], C^3 with encryption disabled, and C^3 with encryption turned on. Results are normalized to a baseline for its execution on unmodified Pin. k was chosen to be the minimum that successfully distributes the basic block for secret sharing, whose values are shown in brackets.

Being consistent with results reported in the original paper [13], ISR does not incur observable slow down compared with execution on unmodified Pin since there is no additional instrumentation. C^3 presents very similar results when encryption is

---

[5]http://code.google.com/p/pyftpdlib.

[6]http://docs.libmemcached.org/bin/memslap.html.

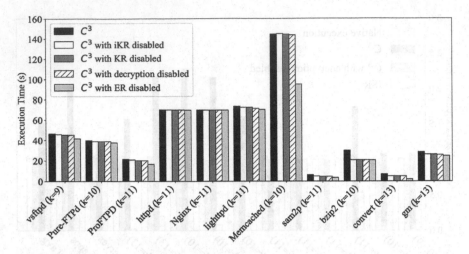

Fig. 4.8 Detailed overhead of C^3

disabled for the same reason. With encryption turned on, C^3 experiences less than 10% overhead for server applications while non-server applications generally suffer from significantly higher overhead. Note that when compared with their respective native executions, several server applications on C^3 have very small runtime performance, although the average runtime overhead is about 70%.

To gain a better understanding of contributions to such overhead and why non-server applications perform worse, we conduct the next finer-grained analysis of C^3 to see which components of C^3 are the main contributors to the overhead. We first identify the following three main tasks of C^3 that potentially contribute to the performance overhead:

- **Key Reconstruction** (KR). This is performed for every branch in the program, be it a direct branch (whose key reconstruction is denoted as dKR) or an indirect branch (whose key reconstruction is denoted as iKR).
- **Decryption.** Since C^3 uses XOR operation as in ISR [13], decryption incurs minimal overhead as confirmed in Fig. 4.7 in which ISR only results in a small overhead.
- **Execution Redirection** (ER). This happens when execution transitions from unprotected code to protected code. Since it requires saving and restoring the entire register state [33], it could result in significant overhead.

Figure 4.8 shows the overhead of C^3 with certain components disabled to more accurately attribute the overhead to the corresponding components. This time, the overhead is presented in seconds without normalization (to visualize the small differences). We have two important observations.

First, iKR (whose contribution can be seen by comparing the bars for C^3 and those for C^3 with iKR disabled) contributes more overhead than dKR (whose contribution can be seen by comparing the bars for C^3 with iKR disabled and those for C^3 with KR disabled). This is mainly due to optimizations C^3 implements for direct branches,

Table 4.3 Number of various branches executed

Programs	iKR #	dKR #	ER #
vsftpd	3.99×10^6	2.43×10^7	1.38×10^6
Pure-FTPd	1.09×10^6	6.29×10^6	2.49×10^5
ProFTPD	5.07×10^6	5.67×10^7	1.68×10^6
httpd	1.16×10^5	4.82×10^5	9.97×10^4
Nginx	3.43×10^3	2.51×10^5	4.70×10^3
lighttpd	1.75×10^5	2.28×10^6	6.31×10^4
Memcached	2.37×10^7	1.54×10^8	7.64×10^6
sam2p	2.22×10^7	1.33×10^8	1.22×10^5
bzip2	1.36×10^8	7.64×10^9	5.83×10^4
convert	2.55×10^7	6.81×10^7	4.36×10^5
gm	1.14×10^6	4.25×10^7	1.21×10^5

where key reconstruction is done only once and results are cached for subsequent decryption. Such optimization does not apply to indirect branches since the control transfer target changes in each indirect branch. Therefore, applications with more indirect branches suffer higher overhead on C^3.

Table 4.3 records the number of indirect branches, direct branches, and transitions from unprotected to protected code. Note that bzip2 has a larger number of indirect branches executed, which explains its higher overhead on C^3.

Our second observation from Fig. 4.8 is on execution redirection ER. We found that ER contributes significantly to the performance overhead for vsftpd, proftpd, memcached and the non-server applications except bzip2, which can be explained by the numbers in the last column of Table 4.3.

Space Overhead

The redistribution of basic blocks in C^3 makes use of a potentially large address space with gaps among various basic blocks; see Sect. 4.3.3. The resulting size of the binary executable mainly depends on the length of the secret, i.e., k. For example when $k = 12$, the address of an instruction can be as big as 2^{24}.

Figure 4.9 shows the resulting file sizes after C^3 processing with two settings—one using a smallest possible setting of k (which varies among different programs) and the other with $k = 14$. We argue that although the size of the binary increases significantly with bigger values of k, storage is cheap and it is usually not an issue with hard-disk space. That said, a larger k also results in slightly bigger runtime overhead as more instructions are executed to extract the values of x and y from an address, and key reconstruction could also require slightly more instructions executed.

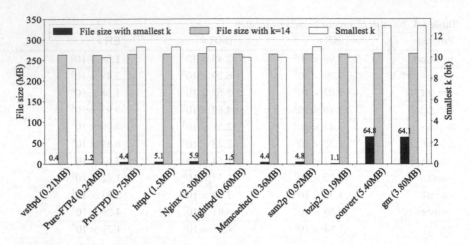

Fig. 4.9 File size with different secret sizes

4.6   Discussion

4.6.1   Return-into-Pin

C^3 provides CFI protection on the application but not the dynamic instrumentation platform, i.e., Pin. An attacker, in theory, could perform an attack by returning into instructions in Pin so that control is diverted directly into the gadgets found in Pin. We call such an attack "return-into-Pin". Such a control transfer would circumvent C^3, enabling the attacker to successfully execute control-flow hijacking attacks. Our design of C^3 is compatible with other isolation hardening solutions, such as Software-based Fault Isolation (SFI) [34], though, which can instrument memory writes to check whether the application attempts to write to a page "owned" by Pin. Another (probably better) defense is to implement the execution environment in a more isolated layer such as the OS layer, the hypervisor layer, the hardware layer, or even inside SGX [16].

That said, once instructions are in Pin's code cache, Pin will not instrument them but jump there directly, which improves the performance of C^3. Meanwhile, such optimization does not hurt security since Pin uses a local hash table for each individual indirect branch transfer, which will contain only the correctly decrypted targets. Any new targets will result in a hash table miss and basic block decryption.

4.6.2 Return-into-libc

In general, CFI does not defend against all return-into-libc attacks. Specifically, C^3 does not encrypt instruction sequences in the .plt section, and so any return instructions can transfer control to entries in the .plt section. In order to protect these library function calls, one could statically compile the libraries into the application.

4.6.3 Length of the Keys

Brute-force attacks have been introduced to reconstructing the encryption keys in ISR [35], which is also applicable to C^3. However, since we use a different key for encrypting each basic block, such brute-forcing will be ineffective because a successful attack typically requires the reconstruction of keys for multiple basic blocks. To this end, we believe that using XOR as the encryption algorithm for improved performance is justifiable, although C^3 can definitely use a more secure encryption scheme. We currently use a 32-bit master key since it is unique for the entire program. C^3 could improve its security with a longer master key of, say, 80 bits.

4.6.4 Fine-Grained CFI Enforcement

C^3 can be extended to enforce fine-grained CFI. For example, C^3 can enforce the fine-grained CFI policy for forward-edge indirect branch transfer instructions with our improved policy described in Sect. 3.5 by classifying functions and indirect call instructions into different clusters according to the number of arguments they can accept, and then encrypting basic blocks with the more accurate set of control transfers derived. We note that enforcing a finer-grained CFI policy could likely reduce the execution time and space overhead of C^3 due to fewer valid control transfers on average and consequently less secret sharing and block redistribution needed.

4.6.5 Other Limitations

First, C^3 relies on static analysis and rewriting of binaries. The current implementation does not support dynamically generated code or self-modifying code.

Second, C^3 prevents attackers from directly reading the code and finding useful gadgets. However, code pointers in data areas such as stack and heap are still vulnerable to indirect memory disclosure. For example, if the protected binary has a format string vulnerability, the attacker can print out the valid memory locations for return instructions, which may allow an attacker to use, e.g., call-preceded gadgets. This is a rather general limitation shared by other techniques performing binary rewriting [1, 2, 28].

Third, C^3 renders caching and pipelining less effective. It is a limitation for most ISR approaches, excluding those performing decryption when there are I-cache misses and store plain text in the I-cache.

Lastly, C^3 requires symbol names in the executable to enable patching function and data pointers after basic block redistribution. It also requires that data and code be completely separated to enforce instruction encryption. For binaries that do not contain symbol information, we can use external tools, e.g., Unstrip[7] and others [36, 37], to restore the symbol information. Similarly, there are approaches to identify data embedded within code [1, 38].

4.7 Summary

We present C^3, a new CFI technique that embeds the CFG into instructions to perform CFI checks without relying on additional data structure like the read-only table used in existing CFI approaches. It encrypts each basic block with a key that can be reconstructed by any of its valid callers with the help of a secret sharing scheme. During execution, C^3 reconstructs the key when a branch transfer instruction is encountered. Our evaluation shows that C^3 can effectively defend against most control-flow hijacking attacks with moderate overhead.

References

1. M. Zhang, R. Sekar, Control flow integrity for cots binaries, in *Proceedings of the 22nd USENIX Security Symposium* (2013), pp. 337–352
2. C. Zhang, T. Wei, Z. Chen, L. Duan, L. Szekeres, S. McCamant, D. Song, W. Zou, Practical control flow integrity and randomization for binary executables, in *Proceedings of the 34th IEEE Symposium on Security and Privacy* (IEEE, 2013), pp. 559–573
3. B. Niu, G. Tan, Modular control-flow integrity, in *Proceedings of the 21st ACM Conference on Computer and Communications Security* (ACM, 2014), pp. 577–587
4. B. Niu, G. Tan, Per-input control-flow integrity, in *Proceedings of the 22nd ACM Conference on Computer and Communications Security* (ACM, 2015), pp. 914–926
5. C. Tice, T. Roeder, P. Collingbourne, S. Checkoway, Ú. Erlingsson, L. Lozano, G. Pike, Enforcing forward-edge control-flow integrity in {GCC} & {LLVM}, in *Proceedings of the 23rd USENIX Security Symposium* (2014), pp. 941–955
6. M. Abadi, M. Budiu, U. Erlingsson, J. Ligatti, Control-flow integrity, in *Proceedings of the 12th ACM Conference on Computer and Communications Security* (ACM, 2005), pp. 340–353
7. M. Zhang, R. Sekar, Control flow and code integrity for COTS binaries: an effective defense against real-world ROP attacks, in *Proceedings of the 31st Annual Computer Security Applications Conference* (2015), pp. 91–100

---

[7]http://paradyn.org/html/tools/unstrip.html.

8. E. Bosman, K. Razavi, H. Bos, C. Giuffrida, Dedup Est Machina: memory deduplication as an advanced exploitation vector, in *Proceedings of the 37th IEEE Symposium on Security and Privacy* (IEEE, 2016), pp. 987–1004

9. H. Hu, S. Shinde, S. Adrian, Z.L. Chua, P. Saxena, Z. Liang, Data-oriented programming: on the expressiveness of non-control data attacks, in *Proceedings of the 37th IEEE Symposium on Security and Privacy* (IEEE, 2016), pp. 969–986

10. G.C. Necula, Proof-carrying code. Design and implementatio, in *Proof and System-Reliability* (Springer, 2002), pp. 261–288

11. G.S. Kc, A.D. Keromytis, V. Prevelakis, Countering code-injection attacks with instruction-set randomization, in *Proceedings of the 10th ACM Conference on Computer and Communications Security* (ACM, 2003), pp. 272–280

12. A. Shamir, How to share a secret. Commun. ACM **22**(11), 612–613 (1979)

13. G. Portokalidis, A.D. Keromytis, Fast and practical instruction-set randomization for commodity systems, in *Proceedings of the 26th Annual Computer Security Applications Conference* (ACM, 2010), pp. 41–48

14. E.G. Barrantes, D.H. Ackley, T.S. Palmer, D. Stefanovic, D.D. Zovi, Randomized instruction set emulation to disrupt binary code injection attacks, in *Proceedings of the 10th ACM Conference on Computer and Communications Security* (ACM, 2003), pp. 281–289

15. A. Papadogiannakis, L. Loutsis, V. Papaefstathiou, S. Ioannidis, ASIST: architectural support for instruction set randomization, in *Proceedings of the 20th ACM Conference on Computer and Communications Security* (ACM, 2013), pp. 981–992

16. Intel Corporation: Intel software guard extensions (Intel SGX) (2019), https://software.intel. com/en-us/sgx/

17. N.A. Quynh, Capstone: Next-gen disassembly framework, *Black Hat USA* (2014)

18. V. Van Der Veen, E. Göktas, M. Contag, A. Pawoloski, X. Chen, S. Rawat, H. Bos, T. Holz, E. Athanasopoulos, C. Giuffrida, A tough call: mitigating advanced code-reuse attacks at the binary level, in *Proceedings of the 37th IEEE Symposium on Security and Privacy* (IEEE, 2016), pp. 934–953

19. A.R. Bernat, B.P. Miller, Anywhere, any-time binary instrumentation, in *Proceedings of the 10th ACM SIGPLAN-SIGSOFT Workshop on Program Analysis for Software Tools* (ACM, 2011), pp. 9–16

20. C.-K. Luk, R. Cohn, R. Muth, H. Patil, A. Klauser, G. Lowney, S. Wallace, V.J. Reddi, K. Hazelwood, Pin: building customized program analysis tools with dynamic instrumentation, in *Proceedings of the 26th ACM Conference on Programming Language Design and Implementation* (ACM, 2005), pp. 190–200

21. D. Bruening, *Efficient, transparent, and comprehensive runtime code manipulation*, Ph.D. thesis, Massachusetts Institute of Technology, 2004

22. E. Göktas, E. Athanasopoulos, H. Bos, G. Portokalidis, Out of control: overcoming control-flow integrity, in *Proceedings of the 35th IEEE Symposium on Security and Privacy* (IEEE, 2014), pp. 575–589

23. L. Davi, A.-R. Sadeghi, D. Lehmann, F. Monrose, Stitching the gadgets: on the ineffectiveness of coarse-grained control-flow integrity protection, in *Proceedings of the 23rd USENIX Security Symposium* (2014)

24. M. Conti, S. Crane, L. Davi, M. Franz, P. Larsen, M. Negro, C. Liebchen, M. Qunaibit, A.-R. Sadeghi, Losing control: on the effectiveness of control-flow integrity under stack attacks, in *Proceedings of the 22nd ACM Conference on Computer and Communications Security* (ACM, 2015), pp. 952–963

25. V. Pappas, M. Polychronakis, A.D. Keromytis, Transparent {ROP} exploit mitigation using indirect branch tracing, in *Proceedings of the 22nd USENIX Security Symposium* (2013), pp. 447–462

26. Y. Cheng, Z. Zhou, Y. Miao, X. Ding, H. Deng, et al., ROPecker: a generic and practical approach for defending against ROP attack, in *Proceedings of the 21th Annual Network and Distributed System Security Symposium* (2014)

27. M. Payer, A. Barresi, T.R. Gross, Fine-grained control-flow integrity through binary hardening, in *Proceedings of the 12th International Conference on Detection of Intrusions and Malware, and Vulnerability Assessment* (Springer, 2015), pp. 144–164
28. M. Wang, H. Yin, A.V. Bhaskar, P. Su, D. Feng, Binary code continent: finer-grained control flow integrity for stripped binaries, in *Proceedings of the 31st Annual Computer Security Applications Conference* (ACM, 2015), pp. 331–340
29. Y. Lin, X. Tang, D. Gao, J. Fu, Control flow integrity enforcement with dynamic code optimization, in *Proceedings of the 19th International Conference on Information Security* (Springer, 2016), pp. 366–385
30. K.Z. Snow, F. Monrose, L. Davi, A. Dmitrienko, C. Liebchen, A.-R. Sadeghi, Just-in-time code reuse: on the effectiveness of fine-grained address space layout randomization, in *Proceedings of the 34th IEEE Symposium on Security and Privacy* (IEEE, 2013), pp. 574–588
31. A. Bittau, A. Belay, A. Mashtizadeh, D. Mazières, D. Boneh, Hacking blind, in *Proceedings of the 35th IEEE Symposium on Security and Privacy* (IEEE, 2014), pp. 227–242
32. N. Carlini, A. Barresi, M. Payer, D. Wagner, T.R. Gross, Control-flow bending: on the effectiveness of control-flow integrity, in *Proceedings of the 24th USENIX Security Symposium* (2015), pp. 161–176
33. H. Pan, K. Asanović, R. Cohn, C.-K. Luk, Controlling program execution through binary instrumentation. ACM SIGARCH Comput. Archit. News **33**(5), 45–50 (2005)
34. R. Wahbe, S. Lucco, T.E. Anderson, S.L. Graham, Efficient software-based fault isolation, vol. 27, no. 5 (1994), pp. 203–216
35. A.N. Sovarel, D. Evans, N. Paul, Where's the feeb? The effectiveness of instruction set randomization, in *Proceedings of the 15th USENIX Security Symposium* (2005)
36. R. Qiao, R. Sekar, Function interface analysis: a principled approach for function recognition in cots binaries, in *Proceedings of the 47th Annual IEEE/IFIP International Conference on Dependable Systems and Networks* (2017), pp. 201–212
37. E.C.R. Shin, D. Song, R. Moazzezi, Recognizing functions in binaries with neural networks, in *Proceedings of the 24th USENIX Security Symposium* (2015), pp. 611–626
38. M. Zhang, M. Polychronakis, R. Sekar, Protecting COTS binaries from disclosure-guided code reuse attacks, in *Proceedings of the 33rd Annual Computer Security Applications Conference* (2017), pp. 128–140

Chapter 5
Control-Flow Integrity Enforcement with Dynamic Code Optimization

5.1 Introduction

Prior to the introduction of CFI in 2005, there have already been a lot of research on dynamic code optimization to improve performance of dynamic program interpreters. For example, Wiggins/Redstone [1], Dynamo [2], Mojo [3], and DynamoRIO [4]. Although most of these were not proposed by the security community, there is at least one noticeable work called *program shepherding* [5] which makes use of a general purpose dynamic optimizer RIO [4] to enforce security policies. DynamoRIO and program shepherding provide nice interfaces for enforcing security policies on control transfers, which makes us believe that they can be good candidate architectures for CFI enforcement. Since these well established and mature dynamic code optimizers are proven to introduce minimal overhead, we believe that they could result in a system that significantly outperforms existing CFI implementations.

In this chapter, we present *DynCFI* that enforces a set of security policies on top of DynamoRIO for CFI properties. We detail how this set of policies are designed and implemented, and show that *DynCFI* achieves similar security properties when compared to a number of existing CFI implementations while experiencing a much lower performance overhead of 14.8% as opposed to 28.6% of *BinCFI*. We stress that *DynCFI* is not necessarily an CFI enforcement implementation that has the lowest performance overhead. Instead, our contribution lies on the utilization of the dynamic code optimization system which is a matured system proposed and well studied before CFI was even introduced, and to the best of our knowledge, *DynCFI* is the first implementation of CFI enforcement on top of a dynamic code optimizer.

In the second half of this chapter, we further investigate the exact contribution to this performance improvement. We propose a three-dimensional design space and perform comprehensive experiments to evaluate the contribution of each axis in the design space in terms of performance overhead. Among many interesting findings, we show that traces in the dynamic optimizer, which consist of cached basic blocks stitched together, had contributed the most performance improvement. Results show that traces have decreased the performance overhead from 22.7% to

Y. Lin, *Novel Techniques in Recovering, Embedding, and Enforcing Policies for Control-Flow Integrity*, Information Security and Cryptography, https://doi.org/10.1007/978-3-030-73141-0_5

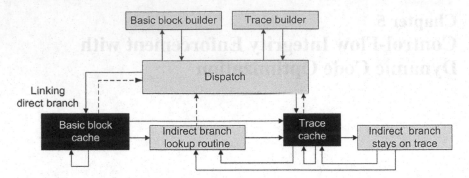

Fig. 5.1   Overview of *DynamoRIO*

14.8%. We also evaluate how branch prediction and indirect branch lookup have changed the performance. To the best of our knowledge, this is the first comprehensive evaluation on the performance overhead contributed by various components of the system, and we believe that this detailed understanding would aid future research and development of efficient CFI enforcement systems.

5.2   Design, Implementation, and Security Comparison

Our objective is to design a practical and efficient CFI enforcement without the extra requirement of recompilation or dependency on debug information. In this section, we first present the design of DynCFI that can be effectively enforced on DynamoRIO and the implementation of it, and then compare the security property it achieves with some existing CFI (and related defense) approaches. Before introducing the details of *DynCFI*, we first overview the workflow of DynamoRIO.

5.2.1   *DynamoRIO*

Figure 5.1 shows an overview of *DynamoRIO* [4], with darker shading indicating the application code to be monitored.

DynamoRIO first copies basic blocks into the basic block cache. If a target basic block is present in the code cache and is targeted via a direct branch, *DynamoRIO* links the two blocks together with a direct jump. If the basic block is targeted via an indirect branch, *DynamoRIO* goes to the indirect branch lookup routine to translate its target address to the code cache address. Basic blocks that are frequently executed in a sequence are stitched together into the trace cache. When connecting beyond a basic block that ends in an indirect branch, a check is inserted to ensure that the

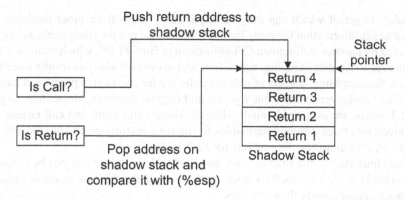

Fig. 5.2 Shadow stack operations

actual target of the branch will stay on the trace. If the check fails, it will go to the indirect branch lookup routine to find the translated address.

To make itself a secure platform on which programs are executed, *DynamoRIO* splits the user-space address into two modes: the untrusted application mode and the trusted and protected RIO mode. This design protects *DynamoRIO* against memory corruption attacks. Meanwhile, the beauty of *DynamoRIO* (and the corresponding good performance) come mainly from the indirect branch lookup which is very efficient in determining control transfer targets with a hashtable. This hashtable maps the original target addresses with addresses in the basic block cache and trace cache so that most control transfers require minimal processing.

5.2.2 Returns

The most frequently executed indirect control transfer instructions are returns. DynCFI maintains a shadow call stack for each thread to remember caller information and the corresponding return address. The whole process is shown in Fig. 5.2. For a call instruction, we store the return address on our shadow stack. For a return instruction, we check whether the address on the shadow stack equals to the address stored at the stack memory specified by %esp. Such a shadow stack enables DynCFI to apply a strict policy that only returning to the caller is allowed, although a relaxed version could also be applied to reduce overhead.

5.2.3 Indirect Jumps and Indirect Calls

We further classify indirect jumps into normal indirect jumps and PLT jumps, such as `jmp offset (base_register)`, which are used to call functions in other modules, target of which can only be exported symbols from other modules. To obtain target information for every indirect branch, we use the static analysis engine provided by another well-known CFI enforcement *BinCFI* [6], which combines linear and recursive disassembling techniques and uses static analysis results to ensure correct disassembling. Targets of indirect calls are function entry points and targets of indirect jumps are function entry points and targets of returns. Meanwhile, targets of PLT jumps are exported symbol address. These valid jump and call targets are organized into three different hashtables to improve performance—one for indirect jumps, one for indirect calls, and one for PLT jumps.

Most importantly, the shadow stack and hashtables we used can just be readable and writable in the DynamoRIO mode, in the user mode, they are readable only, so attackers cannot modify their contents.

5.2.4   Implementation

As discussed in Sect. 5.2.1, the indirect branch lookup routine in *DynamoRIO* maintains a hashtable that maps original control transfer target addresses with addresses of code caches. The hashtable has to be built when the control transfer occurs the first time though. This process, together with the dispatcher which is invoked when matches are not found in the hashtable (see Fig. 5.1), become the natural place of our CFI enforcement, since CFI mainly concerns control transfer targets.

We obtained the source code of *DynamoRIO* version 5.0.0 from the developer's website[1] and added about 700 lines of code (in C) to implement *DynCFI*. Most of the additional code is added to the dispatcher where checks of control-flow transfers are performed. Some code is also added to basic block cache building to implement our shadow call stack and to initialize *DynamoRIO* to load the valid jump/call target addresses into our own hashtables.

DynCFI does not implement the full sets of CFI properties originally proposed by Abadi et al. [7]. In particular, we only perform checks on indirect control transfers at the first time when the target of an indirect branch occurs. However, it does not really impact security, and it is exactly the reason why *DynamoRIO* is widely accepted as an efficient dynamic optimizer—original code is cached in short sequences and security policies, if any, need only be checked the first time the code cache is executed [4]. Subsequent executions of the same code cache will be allowed (without checking) as long as the control transfer targets remain unchanged. Any violations to our policy will miss the (very efficient) indirect branch hashtable lookup and go back to the dynamic interpreter which will consider the control transfer a first timer and perform all the checks (inefficient).

---

[1] http://www.dynamorio.org/.

Table 5.1 Security comparison with other CFI and ROP defenses

Approach	Policy			
	Return	Indirect jump	Indirect call	PLT jump
BinCFI [6]	Call-preceded	Function entry, return address	Function entry	Exported symbol address
CCFIR [8]	Corresponding springboard section			Nil
CFIMon [9]	Call-preceded	Any address in the training set	Any function entry	Nil
ROPdefender [10]	Caller	Nil	Nil	Nil
kBouncer [11]	Call-preceded	Nil	Nil	Nil
LockDown [12]	Caller	Function entry, instruction in the current function	Function entry	Nil
DynCFI	First execution: Caller, Others: Call-preceded	Function entry, return address	Function entry	Exported symbol address

5.2.5 Security Comparison

DynCFI provides comparable security properties with most existing CFI implementation and ROP defense solutions. Table 5.1 shows *DynCFI* (last row) when compared to some of these other approaches.

A caveat here is that in order to improve performance, we make use of the shadow call stack information only when a new target is added to the hashtable (i.e., not checking the shadow call stack if the target address is found in the hashtable). This will make the policy effectively call-proceeded only. Since call-proceeded policy is widely considered as adequate by many other approaches, we apply this performance improvement in our subsequent evaluation. This relaxed policy also enables a fair comparison between *DynCFI* and other CFI enforcement schemes since many others also use a call-proceeded policy.

DynCFI achieves similar security when compared with these existing approaches. In particular, *DynCFI* is mostly comparable to *BinCFI* in that both maintain a list of valid target addresses to be checked at runtime, with one noticeable difference in the enforcement mechanism: *BinCFI* enforces the policies with static instrumentation to translate indirect target address while *DynCFI* uses *DynamoRIO* as the interpreter platform. This makes *BinCFI* the perfect candidate for performance overhead comparison with *DynCFI*, which is the topic of our next Section.

5.3   Detailed Performance Profiling

In this section, we conduct a comprehensive set of experiments on the performance overhead of *DynCFI*. Besides the overall performance overhead, we run some detailed performance profiling to find out the contribution to such overhead by various components of the dynamic optimizer. We wish that such a detailed profiling could shed light on the part that contributes most to the performance overhead, and give guidance to future research in further improvement.

To better understand our evaluation strategy, we present our first attempt in the profiling, show the results, and explain the limitation of this attempt. We then choose an existing CFI implementation for the detailed comparison with *DynCFI*. We analyze the design space of CFI enforcement implementation and organize it along three axes on which the two systems under comparison could be clearly identified. Lastly, we perform a sequence of experiments by modifying individual components of *DynCFI* so that the contribution of each to performance overhead can be evaluated.

5.3.1   Target Applications

To evaluate the performance overhead, we need to subject *DynCFI* (and another CFI implementation for comparison purposes) to some applications. To enable fair comparison with existing work, we used twelve pure C/C++ programs we can find in SPEC CPU2006, which are also used in the evaluation of the original work of *BinCFI* [6], as our benchmarking suite.

Experiments were executed on a desktop computer with an i7 4510u CPU and 8 GB of memory running x86 version of Ubuntu 12.04. Each individual experiment was conducted 10 times, average of which is reported in this paper.

5.3.2   First Attempt in Performance Profiling

As an initial attempt to understand the performance overhead contributed by various components of *DynCFI*, we use program counter sampling to record the amount of time spent in various components of *DynCFI*. We use the ITIMER_VIRTUAL timer which counts down only when the process is executing and delivers a signal when it expires. The handler used for this signal records the program counter of the process at the time the signal is delivered. We sample the program counter every ten milliseconds.

Table 5.2 shows the percentage of time each application spends in various steps in *DynCFI*. It suggests that more than 90% of the time is spent on the application's code on average. Other non-negligible processes include Indirect Branch Lookup

Table 5.2 Percentage of time spent on various components

Application	Application code	IBL inlined	IBL not inlined	BB building	Trace building	Dispatch	Others
bzip2	97.99	0.60	0.00	0.20	1.20	0.00	0.00
gcc	86.78	7.46	0.26	0.91	3.42	1.10	0.07
mcf	97.48	0.42	1.26	0.14	0.07	0.14	0.49
gobmk	80.00	1.08	0.00	2.70	11.35	4.86	0.00
sjeng	94.10	5.67	0.11	0.02	0.09	0.02	0.00
libquantum	99.51	0.49	0.00	0.00	0.00	0.00	0.00
omnetpp	84.88	14.50	0.38	0.06	0.15	0.03	0.01
astar	94.36	4.79	0.78	0.00	0.01	0.04	0.01
namd	99.89	0.69	0.00	0.00	0.02	0.00	0.00
soplex	74.21	25.42	0.03	0.10	0.10	0.10	0.02
povray	89.71	6.88	0.82	0.76	1.01	0.76	0.06
lbm	99.99	0.00	0.00	0.00	0.01	0.00	0.00
Average	91.57	5.62	0.30	0.41	1.45	0.59	0.06

Table 5.3 Statistics of different types of control transfers

Application	%Indirect call	%Indirect jump	%Return	%Direct branch	Total (10^8)
bzip2	0.002	0.002	0.774	99.222	28
gcc	0.434	1.958	7.767	89.841	407
mcf	0.001	0.029	5.402	94.568	50
gobmk	0.001	0.027	4.811	95.161	7
sjeng	1.072	2.289	4.718	91.921	1229
libquantum	0.000	0.000	0.242	99.758	7068
omnetpp	1.609	1.763	33.998	62.630	875
astar	1.698	0.049	19.738	78.515	306
namd	0.000	0.008	3.292	96.700	1159
soplex	0.002	0.018	23.239	76.741	731
povray	2.776	0.154	26.279	70.791	81
lbm	0.000	0.017	0.035	99.948	152

(IBL) inlined with the application's code and that not inlined, basic block and trace cache building, as well as the dispatcher.

In an attempt to explain why some applications, e.g., gcc, omnetpp, soplex, and povray, incur larger overhead, we count the number of different control transfers in each application (runtime) and present statistics in Table 5.3. The correlation between the two tables suggests that larger number of control transfers could lead to the higher overhead.

Table 5.4 Time spent in application code

Application	in *DynCFI* (sec)	Natively (sec)	Overhead (%)
bzip2	4.88	4.86	0.41
gcc	60.73	56.25	7.96
mcf	13.91	14.19	−1.97
gobmk	1.48	1.35	9.62
sjeng	158.93	150.01	5.95
libquantum	813.12	821.63	−1.04
omnetpp	138.54	122.23	13.34
astar	76.16	75.44	0.95
namd	735.51	733.73	0.24
soplex	64.81	61.15	5.98
povray	14.21	14.12	0.64
lbm	375.45	388.14	−3.27

Although it sounds like we have obtained detailed understanding of the performance overhead, there is one important factor that we have overlooked so far—the overhead contribution of the dynamic optimizer on executing the application's code (second column of Table 5.2). In other words, Table 5.2 does not tell us if the dynamic optimizer had sped up or slowed down the execution of the application's code, and what had contributed to that speedup or slowdown. Our further comparison verifies this suspicion, see Table 5.4, as there is noticeable difference in the amount of time spent.

Therefore, we want to further investigate the contribution of various components of the dynamic optimizer in speeding up or slowing down the application's code. We present our second attempt in the rest of this section.

With the objective of finding out contributions to the performance overhead by individual components of the dynamic optimizer, our strategy is to

1. Find an existing CFI implementation X for comparison.
2. Continuously disable or modify individual components of C^3 so that the modified system eventually becomes similar to the implementation of X.
3. In every step of disabling or modifying the components, perform experiments to find the corresponding (difference in) performance overhead.

5.3.3   Picking BinCFI for Detailed Comparison

With this strategy, it is important that we choose an X that

- Is an independent, state-of-the-art implementation of CFI enforcement;

- Shares the same high-level idea with *DynCFI* while validating control transfers with a different approach (e.g., by binary instrumentation) from that of the dynamic optimizer/interpreter as in *DynCFI*.

so that our evaluation could attribute the difference in performance overhead to the dynamic optimizer.

BinCFI and *DynCFI* are similar in that both maintain a set of valid control transfer targets and use a centralized validation routine for CFI enforcement. In both cases, the validation routine maintains a hashtable for the valid control transfer targets.

The difference between *BinCFI* and *DynCFI* is that *BinCFI* obtains the valid target addresses of indirect branches statically and records their corresponding instrumented target addresses into the hashtable, and then replaces the indirect instructions with a direct jump to the CFI validation routine. *BinCFI* satisfies our requirements for the performance comparison, and is therefore chosen for our subsequent detailed evaluation.

5.3.4 Overall Comparison and the Design Space

The overall performance overhead of executing the benchmarking applications under (original, unmodified) *DynamoRIO*, *DynCFI*, and (original, unmodified) *BinCFI* is shown in Fig. 5.3. Results are shown in terms of percentage overhead beyond natively executing the applications on an unmodified Linux Ubuntu system. We obtained the source code implementation of *BinCFI* [6] from its authors.

An interesting observation is that the original *DynamoRIO* and *DynCFI* do not differ much in terms of overhead (a relatively small 1.3% difference). This shows that the interfaces provided by *DynamoRIO* are convenient and effective for CFI enforcement, which confirms our intuition since *DynamoRIO* intercepts all control transfers and no additional intercepting is needed in our modification to *DynamoRIO*.

DynCFI experiences a significantly smaller overhead of 14.8% compared to *BinCFI* at 28.6%. This suggests that the dynamic optimizer provides a more efficient platform for CFI enforcement compared to existing approaches like binary instrumentation as in *BinCFI*. That said, the two systems differ in other aspects and therefore this overall evaluation result is insufficient in attributing the majority of the performance gain to mechanisms of the dynamic optimizer.

As discussed in Sect. 5.3.3, our strategy to this difficulty is to continuously disable or modify individual components of C^3 so that eventually it becomes similar to *BinCFI*, in terms of their operating mechanism as well as the performance overhead. By doing so, we would likely observe degradation of performance (increase in overhead) of the modified system which is definitely due to the corresponding feature disabled or modified. The question is – which individual component or feature to disable or modified?

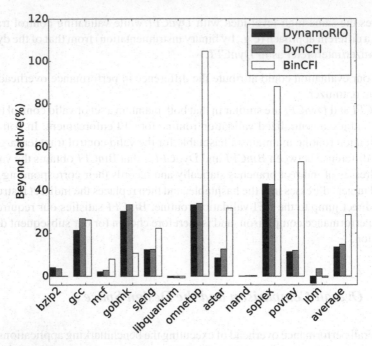

Fig. 5.3 Overall performance overhead

To answer this question, we analyze the internal validation mechanisms of the two approaches and identify three main factors that could significantly contribute to the different performance overhead.

1. **Trace** Trace is the most important mechanism in *DynamoRIO* to speed up indirect transfers. Traces are formed by stitching together basic blocks that are frequently executed in a sequence. Benefits include avoiding indirect branch lookups by inlining a popular target of an indirect branch into a trace (with a check to ensure that the target stays on the trace and otherwise fall back to the full security check), eliminating inter-block branches, and helping branch prediction. Trace is unique in *DynamoRIO* and is not in *BinCFI*.
2. **Branch prediction** Modern processors maintain buffers for branch prediction, e.g., Branch Target Buffer (BTB) and Return Stack Buffer (RSB). The effectiveness of these predictors could get seriously affected due to the modifications to the control transfers. For example, turning a return instruction into a indirect jump would make RSB useless in the branch prediction, potentially leading to an increase in the performance overhead.
3. **Indirect branch lookup routine** Besides implementation details that are not necessarily due to the architectural design (to be discussed more in Sect. 5.3.5), a dynamic optimizer could use a single lookup routine for the entire application including the dynamically loaded libraries, while systems that apply static analysis and binary instrumentation would likely have to use a dedicated lookup routine

for each module because some dynamically loaded libraries might not have been statically analyzed or instrumented. This could contribute to noticeable differences in performance overhead.

We want to explore details into these three axes to see how each of them affects the performance overhead. Other factors that might contribute to the overhead in C^3 which we do not further investigate include

- Building basic block caches;
- Building trace caches;
- Inserting new entries into hashtables;
- Context switches between DynamoRIO and code caches.

5.3.5 Profiling Along the Three Axes

With identification of the three axes, we make our second attempt in detailed understanding of the performance overhead of the two systems. Since executing on *DynCFI* and executing on the original unmodified *DynamoRIO* experience about the same overhead (see Fig. 5.3), our subsequent experiments will only focus on comparing *DynCFI* and *BinCFI*. Also recall that our strategy is to disable or modify one component of *DynCFI* at a time and observe the corresponding change in performance overhead.

Traces
Traces are unique in dynamic optimizers like *DynamoRIO* and *DynCFI*. There are potentially two ways in which traces impact the performance overhead. First, the stitching of basic blocks together eliminates some inter-block branches. Second, each trace has inlined code to check if the control transfer target is still on the trace (we call this InT). If the target is still on the trace, execution will just carry on without further checking; otherwise, a second inlined code (we call this InH) is executed to perform hashtable lookup without collisions. If collision happens, execution will go to the full indirect branch lookup routine (denoted as R). We examine contribution of InT and InH by disabling them individually. We also examine the effect of traces overall and present the results in Fig. 5.4.

Figure 5.4 shows that the contribution due to InT is big, averaging to 5.5%. Exceptions go to bzip2 and soplex which do not gain much with InT mainly because the fall-back of InH is very effective on them (which can be verified from the next-to-zero time spent in IBL not inlined in Table 5.2).

Although performance overhead increases when disabling InT (see Fig. 5.4), *DynCFI* is still better than *BinCFI*. When disabling traces altogether, the overhead of *DynCFI* increases from 14.8% to 22.7% on average, with some going over the overhead in *BinCFI*. This shows that traces are contributing significantly in the low

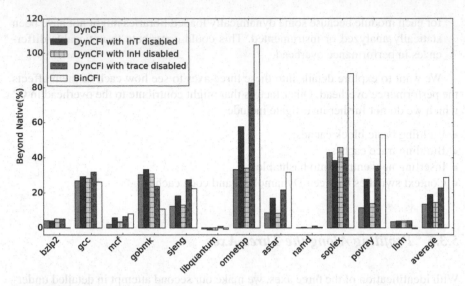

Fig. 5.4 Impact of trace on overhead

Table 5.5 Execution of indirect control transfers

Original transfer		Return	Indirect call/jump
C³	Basic block cache	Jump to R, indirect jump to target	
	Trace cache	InT or InH or jump to R, indirect jump to target	
BinCFI		Return	jump to R, indirect jump to target

overhead of *DynCFI*. For applications with a large percentage of indirect branches (see Table 5.3), *DynCFI* with traces disabled still outperforms *BinCFI*. This suggests that there are other contributing factors in *DynCFI* which we have not evaluated.

Branch Prediction

The way in which *DynCFI* and *BinCFI* intercept and deliver control flow transfers has an implicit effect on branch prediction. Branch prediction is typically achieved by remembering a history of control transfer targets by the same instruction. Both *DynCFI* and *BinCFI* could weaken branch prediction due to R using the same instruction (an indirect jump) to execute control transfers originally executed by different instructions in the application [4, 6]. Table 5.5 summaries how indirect control transfers in an application are executed in *DynCFI* and *BinCFI*.

In summary, *DynCFI* leads *BinCFI* in retaining branch prediction for indirect calls and jumps when trace caches are used due to InT and InH; however, *BinCFI* would perform better than *DynCFI* for returns. That said, note that there are typically far more return instructions than indirect calls and jumps executed for all the applications in our benchmarking suite, see Table 5.3.

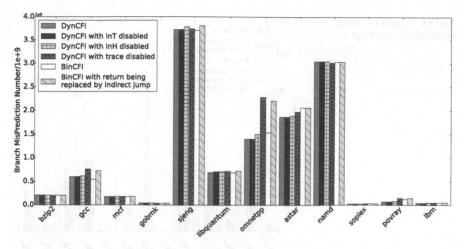

Fig. 5.5 Impact of traces on the number of branch mispredictions

To better understand the effect of various components of *DynCFI* and *BinCFI* on branch prediction, we count the number of mispredictions when executing the benchmarking applications on a number of different settings—*DynCFI*, *DynCFI* with InT disabled, *DynCFI* with InH disabled, *DynCFI* with traces disabled, *BinCFI*, *BinCFI* with returns being replaced by jumps to R, and present the results in Fig. 5.5.

We observe that disabling InH has a larger impact on branch prediction than disabling InT in general. This shows that the inlined hashtable lookup has its fair share of its contribution on lower overhead. It also indirectly shows that the hashtable implementation in *DynCFI* is good in that collisions do not happen often (since R not inlined is not executed often as shown in Table 5.2). Another interesting finding is that replacing returns with indirect jumps on *BinCFI* adds large number of mispredictions for some programs. In terms of overhead, this translates to about 2% more in the overhead as shown in Fig. 5.6.

Indirect Branch Lookup Routine R

The indirect branch lookup routine in *DynCFI* and *BinCFI* very much shares the same strategy. Both use an efficient implementation of a hashtable to record valid control transfer targets. One noticeable difference, though, is that *BinCFI* requires an extra step to check if the target resides within the same software module before directing control to the corresponding R. Each software module has to implement its own copy of R because some dynamically loaded libraries might not have been statically analyzed or instrumented and *BinCFI* cannot use a centralized R for all modules.

Fig. 5.6 Impact of branch
prediction on overhead

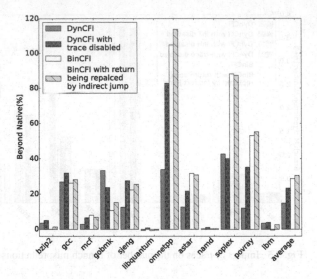

On the other hand, *DynCFI* executes the application on top of a dynamic interpreter without static analysis or binary instrumentation, and therefore has three centralized R (one for returns, one for indirect jumps, and one for indirect calls) for
all software modules. This architectural difference contributes to some additional
performance overhead to *BinCFI*.

Besides the difference due to the architectural design, there are also lower
level differences in implementing R between *DynCFI* (inheriting the same R from
DynamoRIO) and *BinCFI*. In particular, they differ in the indirect jump instructions
used (*DynCFI* uses a register to specify the target while *BinCFI* uses a memory),
the number of registers used throughout the algorithm (and as a result the number of
registers to be saved and restored), and efficiency of the hashtable lookup algorithm.

To evaluate the contribution of R in the overall performance overhead, we replace
R in both *DynCFI* (with traces disabled) and *BinCFI* (with returned replaced with
indirect jumps) with R′, our (supposedly more efficient) implementation of the algorithm, and show the resulting performance overhead in Fig. 5.7.

Comparing these results with those shown in Fig. 5.6, we find that such low-level
details in the implementation of R translates to significant differences in the overhead.
In particular, the difference between C^3 and *BinCFI* shrinks with R′ replacing R,
indicating that the original R used in *DynCFI* is more efficient than that in *BinCFI*.

Fig. 5.7 Performance overhead with unified R′

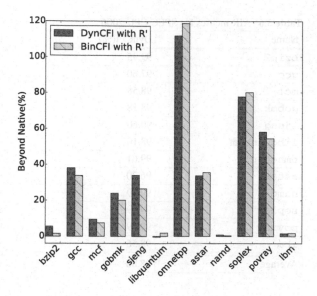

5.4 Security Evaluation and Discussion

5.4.1 Real World Exploits

We use a publicly available intrusion prevention evaluator RIPE [13] to verify that *DynCFI* offers comparable security properties with existing CFI proposals (as analysis presented in Sect. 5.2.5). In particular, we check if *DynCFI* can detect exploits that employ the advanced Return-Oriented Programming (ROP) techniques.

RIPE contains 140 return-to-libc exploits out of which 60 exploit return instructions and 80 exploit indirect call instructions. For the 60 exploits on return instructions, our experiments confirm that *DynCFI* manages to detect all of them because they violate the call-preceded policy we enforced on return instructions.

DynCFI and *BinCFI* share the weakness in detecting exploits that change the value of a function pointer to a valid entry point of a function. Such attacks cannot be detected by most other CFI implementations either [9].

RIPE also contains 10 ROP attacks using return instructions, which are all successfully detected by *DynCFI* as the targets of these gadgets are not call-preceded.

5.4.2 Average Indirect Target Reduction

Zhang and Sekar [6] propose a metric for measuring the strength of CFI called Average Indirect target Reduction (AIR). As *DynCFI* uses different policy on return branches, we apply the same metric to test *DynCFI* when applied to the SPEC

Table 5.6 AIR metrics for SPEC CPU 2006

Name	DynCFI(%)	BinCFI(%)
bzip2	99.95	99.37
gcc	97.60	98.34
mcf	98.58	99.25
gobmk	98.18	99.20
sjeng	99.60	99.10
libquantum	98.10	98.89
omnetpp	99.61	97.68
astar	96.70	98.95
namd	99.99	99.59
soplex	99.49	98.86
povray	99.19	98.67
lbm	98.56	99.46
Average	98.80	98.86

benchmarking suite. Table 5.6 compares the AIR metrics for *DynCFI* and *BinCFI*.
We can find that average AIR for *DynCFI* is 98.80% which is comparable to 98.86%
for the case of *BinCFI*.

5.4.3 Shadow Stack in Full Enforcement

As described in Sect. 5.2, in order to improve the performance, we do not check the
shadow call stack if the target address is found in our hashtable (in which all addresses
have already been fully checked when they were first added to the hashtable).

We understand that a full enforcement of the shadow call stack is more secure
as it ensures that every return jumps to its caller; however, its high performance
overhead is also well documented in previous research [10, 14]. To verify such high
performance overhead, we modify C^3 to check the shadow call stack for every return
instruction, and show the results in Fig. 5.8.

Figure 5.8 shows that C^3 with full enforcement of the shadow stack runs with an
average performance overhead of 29.8%, a big jump from our optimized implementation at 14.8%. Although such a full enforcement of the shadow stack takes away
the performance advantage of C^3 compared to *BinCFI*, C^3 now offers much better
security.

First, we check the AIR metric and find that AIR for C^3 with full enforcement of
the shadow stack increases from 98.80% to 99.66% for SPEC CPU2006, which is
better than that of *BinCFI* at 98.86%.

Fig. 5.8 Performance
overhead with shadow stack

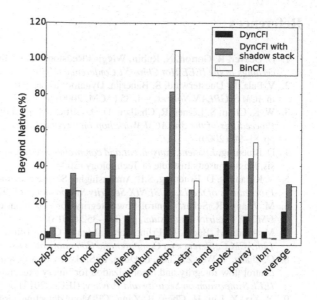

Second, our experiments show that C^3 can now detect some more advanced ROP attacks, e.g., the ROP attack constructed by Goktas et al. [15] using call-preceded gadgets. A call-proceeded-only policy, e.g., that used in *BinCFI*, would miss such advanced attacks.

5.4.4 Fine-Grained CFI Enforcement

DynCFI can be extended to enforce fine-grained CFI. For example, *DynCFI* can enforce the fine-grained CFI policy for forward-edge indirect branch transfer instructions with our improved policy described in Sect. 3.5 by classifying functions into different clusters according to the number of arguments they can accept, and then store them into different lookup tables.

5.5 Summary

In this chapter, we present *DynCFI*, a new implementation of CFI properties on top of a well-studied dynamic code optimization platform. We show that *DynCFI* achieves comparable CFI security properties with many existing CFI proposals while enjoying much lower performance overhead of 14.8% on average compared to that of a state-of-the-art CFI implementation *BinCFI* at 28.6%. Our detailed profiling of *DynCFI* shows that traces, a mechanism in the dynamic code optimization platform, contribute the most to such performance improvement.

References

1. D. Deaver, R. Gorton, N. Rubin, Wiggins/Redstone: an on-line program specializer, in *Proceedings of the IEEE Hot Chips XI Conference* (1999)
2. V. Bala, E. Duesterwald, S. Banerjia, Dynamo: a transparent dynamic optimization system, in *ACM SIGPLAN Notices*, vol. 35 (ACM, 2000) pp. 1–12
3. W.-K. Chen, S. Lerner, R. Chaiken, D.M. Gillies, Mojo: a dynamic optimization system, in *Proceedings of the 3rd ACM Workshop on Feedback-Directed and Dynamic Optimization*, pp. 81–90 (2000)
4. D. Bruening, *Efficient,transparent,and comprehensive runtime code manipulation*. Ph.D. thesis, Massachusetts Institute of Technology (2004)
5. V. Kiriansky, D. Bruening, S.P. Amarasinghe, Secure execution via program shepherding, in *Proceedings of the 11st USENIX Security Symposium*, vol. 92 (2002)
6. M. Zhang, R. Sekar, Control flow integrity for cots binaries, in *Proceedings of the 22nd USENIX Security Symposium*, pp. 337–352 (2013)
7. M. Abadi, M. Budiu, U. Erlingsson, J. Ligatti, Control-flow integrity, in *Proceedings of the 12th ACM Conference on Computer and Communications security* (ACM, 2005) pp. 340–353
8. C. Zhang, T. Wei, Z. Chen, L. Duan, L. Szekeres, S. McCamant, D. Song, W. Zou, Practical control flow integrity and randomization for binary executables, in *Proceedings of the 34th IEEE Symposium on Security and Privacy*, (IEEE, 2013) pp. 559–573
9. Y. Xia, Y. Liu, H. Chen, B. Zang, CFIMon: detecting violation of control flow integrity using performance counters, in *Proceedings of the 42nd Annual IEEE/IFIP International Conference on Dependable Systems and Networks* (IEEE, 2012) pp. 1–12
10. L. Davi, A.-R. Sadeghi, M. Winandy, ROPdefender: a detection tool to defend against return-oriented programming attacks, in *Proceedings of the 6th ACM Symposium on Information, Computer and Communications Security*, (ACM, 2011) pp. 40–51
11. V. Pappas, M. Polychronakis, A.D. Keromytis, Transparent {ROP} exploit mitigation using indirect branch tracing, in *Proceedings of the 22nd USENIX Security Symposium*, pp. 447–462 (2013)
12. M. Payer, A. Barresi, T.R. Gross, Fine-grained control-flow integrity through binary hardening, in *Proceedings of the 12th International Conference on Detection of Intrusions and Malware, and Vulnerability Assessment* (Springer, 2015), pp. 144–164
13. J. Wilander, N. Nikiforakis, Y. Younan, M. Kamkar, W. Joosen, RIPE: runtime intrusion prevention evaluator, in *Proceedings of the 27th Annual Computer Security Applications Conference* (ACM, 2011), pp. 41–50
14. T.H. Dang, P. Maniatis, D. Wagner, The performance cost of shadow stacks and stack canaries, in *Proceedings of the 10th ACM Symposium on Information, Computer and Communications Security*, vol. 15 (2015)
15. E. Göktas, E. Athanasopoulos, H. Bos, G. Portokalidis, Out of control: overcoming control-flow integrity, in *Proceedings of the 35th IEEE Symposium on Security and Privacy* (IEEE, 2014), pp. 575–589

Chapter 6
Conclusions

This book makes contributions in recovering, embedding, and enforcing control-flow integrity policies.

In the first work presented in Chapter 3, we systematically study how compiler optimization would impact function signature recovery with 1,344 real-world applications with various optimization levels, and propose a novel improved mechanism to more accurately recover function signatures. The results show that compiler optimizations have both positive and negative impacts on function signature recovery.

In the second work presented in Chapter 4, we propose C^3, a novel approach to embed CFI policies into instructions rather than consulting read-only tables or inserting tags into the code section compared to other approaches. It embeds the CFI policies and its enforcement into instructions of the program by encrypting each basic block with a key derived from the control-flow graph. This kind of "proof-carrying" code ensures only valid control-flow transfers can decrypt the corresponding instruction sequence and any unintended control-flow transfer would cause program crash. The security evaluation shows that C^3 is able to defend against most control-flow hijacking attacks while suffering from moderate runtime overhead.

The third work presented in Chapter 5 presents *DynCFI*, an efficient way to enforce CFI based on the dynamic code optimization platform DynamoRIO. The result shows that *DynCFI* enjoys much lower performance overhead of 14.8% on average compared to that of a state-of-the-art CFI implementation BinCFI at 28.6%. We further perform comprehensive evaluations and shed light on the exact amount of savings contributed by the various components of the dynamic optimizer including basic block cache, trace cache, branch prediction, and indirect branch loopup.

© The Author(s), under exclusive license to Springer Nature Switzerland AG 2021
Y. Lin, *Novel Techniques in Recovering, Embedding, and Enforcing Policies
for Control-Flow Integrity*, Information Security and Cryptography,
https://doi.org/10.1007/978-3-030-73141-0_6

Printed in the United States
by Baker & Taylor Publisher Services

Printed in the United States
by Baker & Taylor Publisher Services